AMORPHOUS SEMICONDUCTORS

David Adler
Department of Electrical Engineering
Massachusetts Institute of Technology
Cambridge, Massachusetts

published by:

THE CHEMICAL RUBBER CO.
18901 Cranwood Parkway • Cleveland, Ohio 44128

CRC MONOTOPICS SERIES

The primary objective of the CRC Monotopics Series is to provide reference works, each of which represents an authoritative and comprehensive summary of the "state-of-the-art" of a single well-defined scientific subject.

Among the criteria utilized for the selection of the subject are: (1) timeliness; (2) significant recent work within the area of the subject; and (3) recognized need of the scientific community for a critical synthesis and summary of the "state-of–the-art."

The value and authenticity of the contents are assured by utilizing the following carefully structured procedure to produce the final manuscript:

1. The topic is selected and defined by an editor and advisory board, each of whom is a recognized expert in the disciplin.

2. The author, appointed by the editor, is an outstanding authority on the particular topic which is the subject of the publication.

3. The author, utilizing his expertise within the specialized field, selects for critical review the most significant papers of recent publication and provides a synthesis and summary of the "state-of-the-art."

4. The author's manuscript is critically reviewed by a referee who is acknowledged to be equal in expertise in the specialty which is the subject of the work.

5. The editor is charged with the responsibility for final review and approval of the namuscript.

In establishing this new CRC Monotopics Series, CRC has the additional objective of containing the ever-rising cost of publishing, and scientific publishing in particular. By confining the contents of each book to an in-depth treatment of a relatively narrow and well-defined subject and exercising rigorous editorial control, the publishers have ensured that no irrelevant matter is included.

Although well-known as a publisher, CRC now prefers to identify its function in this area as the management and distribution of scientific information, utilizing a variety of formats and media ranging from the conventional printed page to computerized data bases. Within the scope of this framework, the CRC Monotopics Series represents a significant element in the total CRC scientific information service.

B. J. Starkoff, President
THE CHEMICAL RUBBER CO.

CRC
PRESS

This book originally appeared as an article in *CRC Critical Reviews in Environmental Control*, a quarterly journal published by The Chemical Rubber Co. We would like to acknowledge the editorial assistance received by the journals' editors, Prof. Richard G. Bond and Dr. Conrad P. Straub, both at the University of Minnesota. Mr. Harold O. Wyckoff, Armed Forces Radiobiological Research Institute, served as referee for this article.

AUTHOR'S INTRODUCTION

This book grew from a review article prepared for *CRC Critical Reviews in Solid State Sciences.* Between agreeing to write the article in late 1969 and finishing it in June 1971, I was faced with the fact that the quantity of published work in the field was increasing at an exponential rate. Consequently, there were times when it appeared that it would be impossible to complete a comprehensive review, due to lack of convergence. The temptation was strong to paraphrase the beleaguered gas-station attendant of the late fifties: "Turn off the engine – you're gaining on me." Needless to say the review turned out an order of magnitude longer than originally planned.

The field of amorphous semiconductors is one in constant flux. Our ideas are continually changing, and many fundamental concepts still are lacking. The rate of growth of the field shows no signs of saturation, and it is clear that any review cannot remain current for a long period of time. Nevertheless, because of the "pioneering" work in the 1969-71 period, the structure of the problem has been classified, and we now know the relevant parameters which must be both measured and calculated for a wide class of materials in order to test our ideas quantitatively. Thus we have at least passed the fundamental barrier that impedes the evolution of any field of research from an empirical correlation of data to a science. This is comforting, if for no other reason than in the sense that it decreases the probability of embarrassment from any publication in the field, much less a review. But it is well to bear in mind that many quantitative results will change by orders of magnitude and most of the present models will eventually fall by the wayside.

I should like to single out a few people for special thanks in nontechnical matters: my wife, Alice, for putting up with me during the intensive period when the last four chapters were written in a single burst; my secretary, Sally Nutter, for deciphering the chicken scrawls and converting them to a readable albeit massive manuscript; and the Chemical Rubber Company for patiently waiting for the completion of the review and then getting it into print in a time faster than the average Physical Review paper is returned by the referee.

David Adler
Cambridge, Mass.

THE AUTHOR

David Adler is Associate Professor of Electrical Engineering at M.I.T.

Dr. Adler received his B.S. degree from Rensselaer Polytechnic Institute, Troy, New York, and was awarded his Ph.D. degree in physics by Harvard University in 1964. In 1964-65, he was a Research Associate at the United Kingdom Atomic Energy Research Establishment at Harwell, England. He joined the Massachusetts Institute of Technology, Boston, faculty in 1967 after two years as Research Associate.

Dr. Adler's research interests include the electronic properties of low-mobility materials and insulator-metal transitions, in addition to amorphous semiconductors. He has published over 40 papers in technical journals and has presented over 20 invited papers at scientific meetings. He is on the editorial board of the Journal of Nonmetals and is a member of the American Physical Society, the American Vacuum Society, and the American Association of University Professors.

TABLE OF CONTENTS

I. INTRODUCTION

There has been a recent surge of interest in the subject of amorphous semiconductors, a field which only a few years ago was generally considered to be about as scientific as witchcraft or alchemy. The major reason for the change in attitude is not difficult to pinpoint: the turning point was the publication by Ovshinsky[1] detailing the various types of switching phenomena that characterize a large class of amorphous solids, and the subsequent publicity describing many potential applications of these phenomena.[2,3] But, in addition, the field presents several surprising features that stimulate pure academic interest. Firstly, a serious investigation of the properties of amorphous semiconductors requires not only familiarity with one particular area of science or engineering, but rather a detailed knowledge of results from the combined fields of physics, chemistry, metallurgy, and electrical engineering. Secondly, although the study of amorphous semiconductors, especially when compared with that of crystalline materials, is certainly in its infancy, and much fundamental experimental and theoretical work remains to be done, still much sophisticated thought has been devoted to the subject over the past 10 years and great strides have been made in understanding the basic principles. Finally, even a superficial investigation of amorphous semiconductors brings home rather strikingly the fact that solid state physics, even up to the present time, is being taught incorrectly!

The last remark needs some amplification. A text in solid state theory usually defines a solid as a periodic array of atoms, by which is meant that if we choose a particular atom at random, the environment of that atom is precisely the same, no matter how far away we look, as the environment of any other equivalent atom in the material. In this way, a solid can be differentiated from a liquid, in which the short-range environment of a given atom—the nearest and next-nearest neighbors – is usually almost precisely the same as in the solid phase of the same material, but the tenth-nearest or fiftieth-nearest neighbors of the liquid phase are essentially at random with respect to the central atom. Given the definition, a standard text ordinarily proceeds with a classification of the various types of structures compatible with the requirement of equivalent environments—the subject of crystallography. But sooner or later, implications of periodicity are brought home with the proof of Bloch's Theorem, which restricts the class of states available to electrons in a crystalline solid to ones which are delocalized, or spread throughout the entire crystal. From Bloch's Theorem follows the entire energy-band theory of solids, and the explanation of the electrical properties of materials. Energy-band theory can be shown to predict successfully whether a given solid will be a metal, an insulator, or a semiconductor.

It seems that the logic is convincing: periodicity implies Bloch's Theorem, which in turn implies delocalized states, which leads to the observed electrical properties of solids. Clearly then, periodicity is of vital importance to the electrical conductivity of a given material. But this is a premise which can be tested very easily. All we need to do is measure conductivity as a function of temperature, not only at ordinary temperatures, but right up through the melting point into the liquid region. What should we then expect? In most materials, there is a sharp change in density, i.e. in interatomic spacing, at the melting point. Since conductivity is an extremely sensitive function of nearest-neighbor separation, we should anticipate a jump in conductivity purely from this volume discontinuity, even if the crystal structure were preserved through the transition. But, in addition, the long-range periodicity disappears essentially at once at the melting temperature. If periodicity is indeed vital to the electrical properties, we might expect conductivity to change by many orders of magnitude at the transition point.[4]

The actual results of such experiments are easily summarized. With the exception of a handful of materials which change their nearest-neighbor environment upon melting, there is not much change in electrical conductivity at the melting temperature. Metals remain metals, insulators remain insulators, semiconductors remain semiconductors. Typical results are shown in Figure 1. Small discontinuities in conductivity of either sign do exist, but in virtually every case, these can be correlated with discontinuous volume changes, and in a few materials for which no significant volume effects occur (e.g. CdTe), it is impossible to pick out the melting points from the curves of conductivity vs. temperature.

Evidently, the importance of long range periodicity has been overemphasized. We might then ask why, in the face of this evidence, is this myth being sustained. There is indeed an answer to this question: periodicity enables us to solve the problem quantitatively. Without perfect long-range order, we would be faced with solving the quantum-mechanical problem of some 10^{24} mutually interacting electrons moving in the fields of 10^{23} or so mutually interacting ion cores arranged in an irregular pattern, clearly a hopeless proposition. On the other hand, the requirements of equivalent environments permit us to solve the quantum mechanical problem in one primitive cell only, thus bringing about a reduction in mathematical complexity of the order of a factor of 10^{23} to a mathematically feasible task (at least with the aid of high-speed computers).

There is no problem so hopeless that a few general theorems cannot be proved, and even the random lattice has been studied in some detail. In one dimension, the important result can be demonstrated that all electronic states are localized around a particular position in space, rather than extended like the states in a periodic crystal. Various proofs of this result have been given by Mott and Twose[5,6] and by Borland and Bird.[7-9] The physical reason for the localization stems from the fact that an electronic wave function, in general, will be exponentially attenuated as it tunnels through a quantum-mechanical barrier. The extent of the attenuation depends on the phase, but for only an extremely limited range of phase does the wave function fail to decrease at a given barrier. In a completely periodic system, the phase can be chosen such that the wave function falls within the narrow range which avoids attenuation at *every* barrier in the material. These are the Bloch wave functions. On the other hand, one deviation from periodicity in the bulk of the system would, in one dimension, result in continuing exponential decay at each barrier after the defect, and thus a partial localization. Many deviations from periodicity lead to complete localization.

Unfortunately, few useful general results have been rigorously demonstrated as yet for the random lattice in three dimensions, although some speculations have been presented.[6] But, in a major sense, these theorems are irrelevant, since no solid exists in a totally random structure. In fact, as was previously indicated, amorphous materials exhibit a high degree of short-range order. For example, x-ray and electron diffraction results show that the

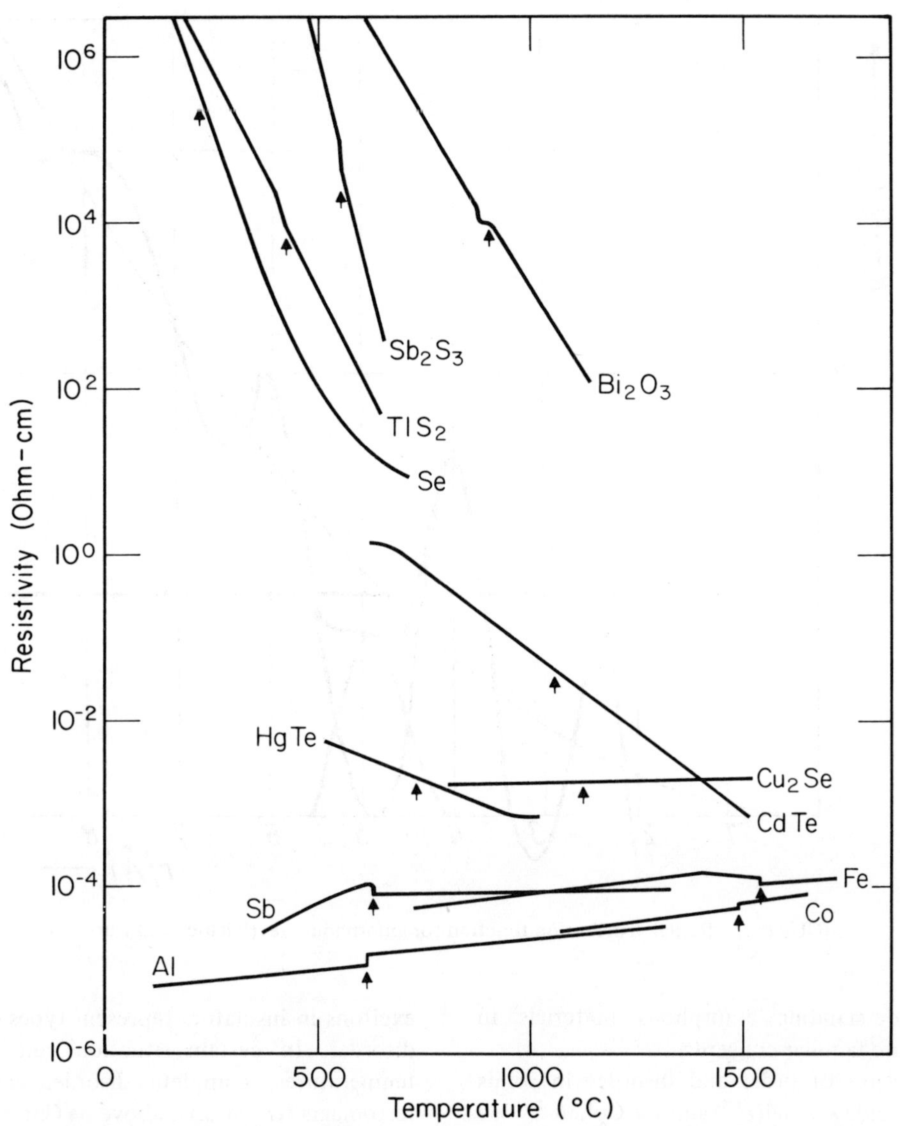

FIGURE 1. Resistivity as a function of temperature for several materials in both the crystalline and liquid states. The melting temperature in each case is indicated by an arrow.

nearest neighbor environments in amorphous Ge and Si are almost precisely the same as in the corresponding crystal.[10,11] On the other hand, beyond about fifth-nearest neighbors, the atoms are distributed nearly uniformly. The short-range order and long-range disorder of amorphous Ge is evident from the experimental radial distribution function shown in Figure 2. Thus we can conclude that a large percentage of the Ge atoms in the amorphous material are surrounded by a regular tetrahedron of nearest-neighboring Ge atoms, with essentially the same separation as in the crystal. From a tight-binding point of view, a good approximation for most semiconductors, the energy levels available to an outer electron in a crystal can be estimated by starting with the atomic energy levels and then perturbing these by introducing the effects of the nearest-neighbor atoms. From this viewpoint, it is understandable that the energy-band structures of crystalline and amorphous Ge could be quite similiar.[12]

A. Order and Disorder in Solids

We have been discussing long-range and short-range order with respect to atomic positions. Other types of order-disorder problems are common in the study of solids, and it is worthwhile to discuss the general subject, with a view

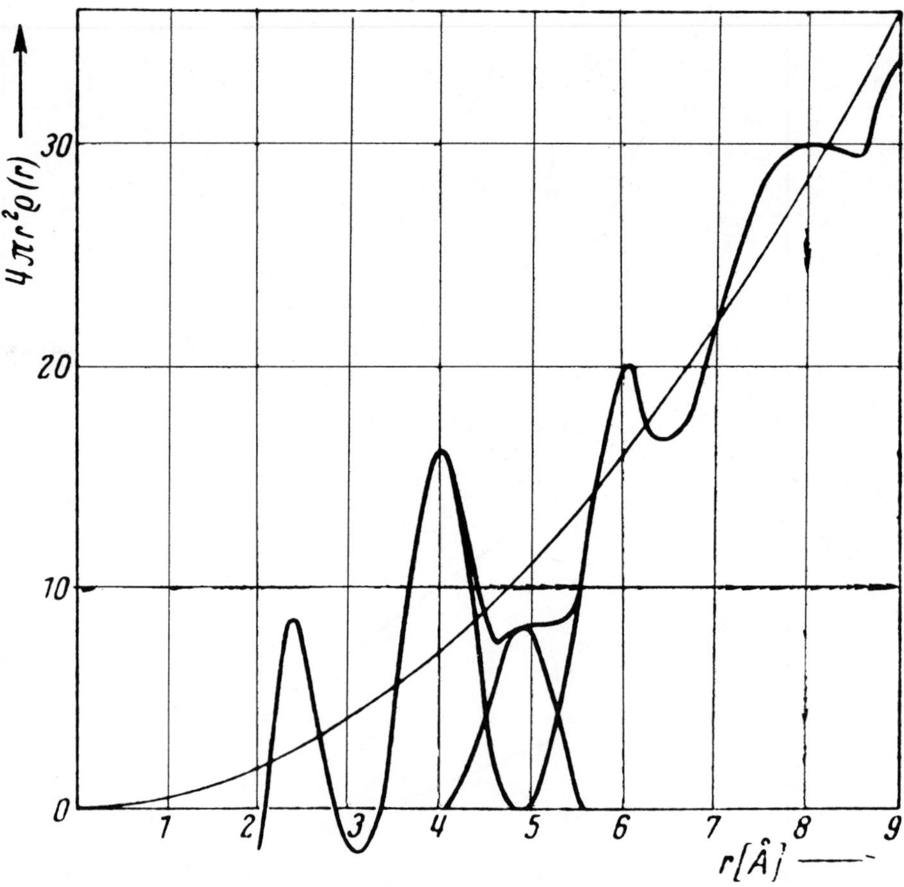

FIGURE 2. Radial distribution function for amorphous Ge (Richter and Furst[10]).

toward understanding amorphous materials in terms of more familiar concepts.

Classification of order and disorder in solids have been made by Seitz[13] and by Cohen.[14] We can distinguish between compositional order, positional order, magnetic (spin) order, and electronic order. A perfect crystal in its ground state has perfect order. Of course, this is never the case, and static imperfections exist. An imperfection in compositional order is an impurity; an imperfection in positional order is a defect; an imperfection in the magnetic order (of systems in which the atoms have a non-vanishing magnetic moment) is a reversed spin; an imperfection in the electronic order is an excited atom. There are also dynamic imperfections, or elementary excitations, which can be associated with all types of order except compositional order. Thus phonons (lattice vibrations) bring about dynamic positional disorder, magnons (spin waves) lead to magnetic disorder, and plasmons (plasma oscillations) in metals and excitons in insulators represent types of electronic disorder. In certain situations, e.g. at elevated temperatures, complete disorder can occur. A ferromagnetic material above its Curie temperature has no long-range magnetic order, although short-range order can persist up to two or three times this temperature. An alloy above its ordering temperature similarly exhibits long-range compositional disorder, but again short-range order. Above the melting point, a material has long-range positional disorder. The same is true in all amorphous solids.

We shall be primarily interested in the electrical properties of amorphous materials, and these are determined by the electronic states. Since the electrons are extremely light compared to the massive ion cores, they quickly adjust to changes of state of the ion cores. (This is the adiabatic approximation,[15] which enables us to separate electronic motion from ion-core motion.) Thus the electronic energy levels are extremely sensitive to variations of composition, ionic position, and spin orientation. In the course of analyzing electronic properties of disordered alloys and spin systems, techniques have been developed for approximating the effects of compositional and spin disorder on electronic states. It has been found that the disordered alloy problem can be handled with a virtual-crystal approximation as a starting point.[16] In this case, the potential energy due to any ion core is represented by $xV_A + (1-x) V_B$, where x is the fractional amount of atom A in the alloy, and V_A and V_B are the potentials of the A and B ion cores, respectively. The virtual-crystal approximation has the major advantage of maintaining the periodicity of the potential energy, so that Bloch's Theorem remains applicable. Similarly, it has recently been suggested[17] that the onset of spin disorder in ferromagnetic or antiferromagnetic materials does not significantly alter the electronic energy bands as calculated with the assumption of perfect order. In such cases, the major effect of spin disorder appears to be an extra scattering mechanism[18] which could limit the mobility of free carriers if phonon and impurity scattering are sufficiently small. Even a general type of positional disorder, that due to the presence of crystal surfaces in all real materials, is handled by the imposition of periodic boundary conditions,[19] thus enabling us to apply Bloch's Theorem. In all of these examples it is evident that the Bloch states calculated from the assumption of a perfectly periodic crystal represent a good basis for the solution to the disordered problem.

B. Classification of Amorphous Solids

Amorphous solids are most easily classified in exactly the same manner as are crystalline solids, i.e. by the type of chemical bonding that is primarily responsible for cohesive energy of the material. In crystalline solids, we consider five major classes—ionic, covalent, metallic, van der Waals, and hydrogen-bonded materials. This classification scheme can also be used for amorphous solids. Amorphous metals do exist, and appear to have approximately the same conductivity as the corresponding crystalline material, a further sign that the effects of long-range disorder on the electronic properties of solids are quantitative rather than qualitative. However, metals do not fall within the scope of a discussion of amorphous semiconductors, and we shall not discuss them further. Van der Waals and hydrogen-bonded solids generally have low cohesive energies and thus low melting temperatures, and consequently the corresponding amorphous solids have not been studied to any great extent.

Thus, the field of amorphous semiconductors can be broken down into ionic and covalent materials. The ionic materials which have been studied most are the halide and oxide glasses, particularly the transition-metal oxide glasses. Recently, simpler amorphous solids, such as V_2O_5 and NiO, have been prepared by rf-sputtering techniques. Being primarily ionic, the composition of these materials cannot be made to vary over a wide range, and the pure materials have just positional disorder. On the other hand, in the transition-metal oxides, the presence of impurities usually results in transition-metal ions of two different valence states. For example, introduction of P_2O_5 into V_2O_5 produces V^{4+} ions as well as V^{5+} ions. Thus, these materials can be said to possess some degree of electronic disorder. Furthermore, the unsaturated transtion-metal ions generally contribute some spin disorder.

It will be convenient to further break the covalent amorphous semiconductors into two classes, by separating out the purely elemental materials. This group, which is perfectly covalently bonded, of course, includes Si, Ge, S, Te, and Se, among others. Since all atoms are necessarily the same, these materials possess only positional disorder. Thus, they are the simplest amorphous semiconductors to investigate theoretically. The remaining covalent materials include such binary materials as As_2Se_3 and GeTe, as well as the multicomponent boride, arsenide, and chalcogenide glasses. Being primarily covalent, there is nothing magic about the chemical composition. As_2Se_3 nominally contains 40% As and 60% Se, but there is no reason why we cannot make an alloy of, say 45% As and 55% Se, or any other proportion we desire. Since the material is amorphous, we are not tied to any crystalline structure, and thus defects, dangling bonds, and antistructure are not problems. Similarly, we can combine As, Se, Ge, and Te, and produce, for example, $As_{31}Se_{21}Ge_{30}Te_{18}$, an example of a chalcogenide glass.[20] The significance of this is that the covalent amorphous semiconductors always possess compositional as well as positional disorder. As we shall see, the profound effects of this joint disorder on the electronic band structure are responsible for the distinctive properties of these glasses, and justifies their separation from the class of elemental amorphous semiconductors.

As in crystalline solids, many mixed ionic-covalent amorphous materials exist, and these have to be analyzed carefully. The similarities in short-range order and bond lengths between corresponding crystalline and amorphous solids imply that the atomic electronegativities are independent of the nature of the material, and we can use these to estimate the average fractional ionicity of a given alloy.

C. Scope of this Review

The same problems which concern researchers investigating crystalline solids are also of interest to those studying amorphous materials. However, the analysis is generally much more complex, both theoretically and experimentally, in the absence of crystallinity and consequently the state of the art is quite primitive. Once the adiabatic approximation[15] is invoked, the problem can be broken down into two parts, the position and motion of the ion cores and the position and motion of the outer electrons. The analysis of the equilibrium positions of the ion cores is the amorphous analogue of crystallography, and it can be studied by the same techniques, e.g. x-ray diffraction, electron diffraction, paramagnetic resonance, as used on crystals. However, the lack of long-range periodicity makes the results of these experiments extremely difficult to interpret, and the question of the structure of amorphous materials is still in doubt. Because of this, the problem of normal modes of vibration of the ion cores about their equilibrium sites, the analogue of lattice dynamics is not yet an important field of investigation. Since we should expect only small deviations of the ion cores from their equilibrium positions, the quasi-particle concept of phonons should eventually prove useful in amorphous solids. On the other hand, the analogy will be far from trivial; for example, we shall have to consider the possibility of localized vibrational modes, we must not expect to discuss E vs. k dispersion relations, and we should be careful about distinguishing between acoustic and optical branches or between longitudinal and transverse modes. Because of the long-range nature of the Coulomb interactions between ion cores, interpretation of the phonon spectrum of crystalline solids in terms of a force-constant model often requires explicit consideration of sixth-nearest neighbors and up.[21] It is reasonable to anticipate that similar results hold for amorphous materials, and, if so, an analysis of ionic vibrations under the assumption that the phonon spectrum is essentially the same as that of the corresponding periodic crystal will almost undoubtedly fail. Although the nearest-neighbor and occasionally even the second-nearest neighbor configuration of the corresponding amorphous and crystalline materials may be the same, the extreme deviations of third and higher neighbors from their proper lattice positions should destroy the applicability of any pseudocrystal models. Alternatively, if the phonon spectrum can be explained on the basis of a nearest-neighbor and second-neighbor force-constant model, the spectra of corresponding amorphous and crystalline solids could be very similar. Since structural problems are only of interest to us insofar as they lead to an understanding of the semiconducting properties of amorphous solids, we shall discuss the results of structural experiments only briefly in the remainder of this review.[22]

The second part of a solid-state problem is a study of the dynamics of the outer electrons of the material, and it is this which will be our primary interest. Although the electronic problem, in the absence of crystal periodicity, cannot be solved in the usual manner, we are helped immensely by the fact that the short-range order is the same in corresponding amorphous and crystalline solids. For one thing, the one-electron approximation is quite good for calculating the energy bands of a large class of crystals. Since the validity of this approximation depends primarily on interatomic spacing,[23] we should expect it to apply equally well to amorphous as to crystalline materials, at least for the extended states. Thus, although we cannot reduce the electronic problem to solving for the one-electron energies as a function of the crystal wave vector **k** in an appropriate Brillouin Zone, and although **k** is not a good quantum number, we still can write down the one-electron Schrodinger equation for an amorphous solid as

$$-\frac{\hbar^2}{2m}\nabla_i^2\,\phi_n(\underset{\sim}{r}_i) + V(\underset{\sim}{r}_i)\,\phi_n(\underset{\sim}{r}_i) = E_n\phi_n(\underset{\sim}{r}_i) \qquad (1)$$

where $V(\mathbf{r}_i)$ is the potential of electron i due to all the ion cores and all the other electrons in the material. Thus, the one-electron energies E_n exist,

* In the equations the symbol ~ appearing under a letter will indicate a vector.

and although we must give up the concept of plotting E(**k**) functions, we can retain the idea of density-of-states diagrams, g(E), the number of one-electron states per unit volume in an infinitesimal energy range between E and E + dE.

Given the density of states, g(E), we are still far from describing the dynamics of the electrons in the solid. Under the usual assumption that the interacting electrons still obey Fermi-Dirac statistics, we can evaluate the average occupancy of these states at all temperatures. But this is just about as far as we can push the analogy with the crystalline problem. For crystalline solids, we have an important result, i.e. Bloch's Theorem, which is invalid in the absence of periodicity. Thus, we cannot prove that the wave functions, $\Phi_n(\mathbf{r}_i)$, are delocalized.

Once again we can be rescued by a more fundamental concept, that of mobility, or electronic velocity per unit applied electric field. The mobility depends on the entire state of the system and, in general, cannot be associated with a given one-electron state. However, we can define the mobility of any conduction-band state as the velocity per unit field attained by an electron placed in the state, under the assumption that all valence band states remain full throughout the conduction process. Similarly, the mobility of a valence-band state can be obtained by placing a single hole in the state and evaluating the velocity per unit field, assuming the conduction band states are completely empty.

For a perfectly periodic crystal, all states are delocalized, and the mobility of any state is infinite. In real crystals, the mobility is limited by scattering arising from deviations from perfect periodicity. In highly disordered systems, intense scattering due to large fluctuations in the one-electron potential can reduce the mobility to quite small values, despite the delocalized nature of the states. Thus, mobility is a more fundamental concept than localization.

As previously indicated, we cannot prove that the eigenstates for amorphous solids are delocalized. As a qualitative guide, we might, in fact, postulate that all the solutions to Schrodinger's Equation 1 are localized. Localized states must have zero mobility if they do not spatially overlap other states with the same energy. However, if a sufficient density of overlapping localized states exists at a given energy, quantum-mechanical tunneling between them can lead to a finite mobility, despite the localized nature of the states. This situation reflects the fact that extended linear combinations of the degenerate localized states can be constructed, and these must also be eigenstates of the same Schrodinger's equation. Thus, the mobility of the states in an amorphous solid sensitively depends on the density of states, g(E). Where the density of states is large, the mobility can be finite; where the density of states is small, the mobility is essentially zero.

In nearly periodic systems, it is ordinarily assumed that the mobility of the extended states depends only on the energy.[24] Such an approximation is not so obvious when finite mobilities result from linear combinations of degenerate localized states, since the possibility exists that some of the eigenstates at a given energy are segregated in a particular region of space and are thus much more localized than other sets at the same energy. However, since quantum-mechanical tunneling between the segregated states and extended states at the same energy always has some probability, we should, in fact, perform a weighted average over the respective mobilities to obtain a single function μ (E).

Once the single particle density of states g(E) and the corresponding mobility μ(E) are known, we can analyze the electrical properties of the material—conductivity, thermoelectric power, Hall effect, magnetoresistance, etc. The same information enables us to predict the optical properties, i.e. absorption, reflectivity, photoemission, photoconductivity, etc. All of these classes of experiments can be performed as a function of temperature, of pressure, and, in the alloys, of composition. It is the result of such measurements which will be of primary interest in this review.

In Section II, we present a brief discussion of the energy-band theory of crystalline solids, paying particular attention to techniques which have been developed for handling the effects of deviation from periodicity. This review is intended primarily to develop several concepts that are important in understanding some of the problems in amorphous semiconductors, and can be omitted by those familiar with modern solid-state theory. In section III, some of the recent developments in the theory of amorphous semiconductors are discussed, despite the fact that a rigorous, quantitative approach is still lacking. This section is intended to be a selective guide to current research directions, rather than a pedagogic basis on which future advancements must rest. However, the models discussed can be used as a structure with which to analyze the experimental results detailed in the bulk of the review. Sections IV, V, and VI, deal with the experimental results on representative materials comprising the three major classes of amorphous semicondutors, the elemental materials (section IV), the covalent materials (section V), and the ionic materials (section VI). A brief review of the several types of switching phenomena which characterize amorphous semiconductors is given in section VII. Some general conclusions are noted in section VIII.

II. Energy-Band Theory of Solids

A. Perfect Crystals

Once the adiabatic approximation is applied, Schrodinger's equation for the electronic problem can be written

$$\sum_i \left[\frac{p_i}{2m} + V(\mathbf{r}_i) + \frac{1}{2}\sum_{j\neq i}{}' \frac{e^2}{|\mathbf{r}_i - \mathbf{r}_j|}\right] \Phi_n(\mathbf{r}_i) = E_n \Phi_n(\mathbf{r}_i) \qquad (2)$$

where $\mathbf{p}_i$ and $\mathbf{r}_i$ are, respectively, the momenta and positions of the ith electron, and $V(\mathbf{r}_i)$ contains both the attractive potential between an electron and all the ion cores as well as the (constant) mutual repulsion of the ion cores. The adiabatic approximation also allows us to solve Equation 2 for the electronic energy levels, E_n, with the ion cores fixed in their equilibrium positions, since the effects of ionic displacement are treated in the lattice problem. For a perfect crystal, the potential energy $V(\mathbf{r}_i)$ must be periodic as well; i.e.

$$V(\mathbf{r}_i + \mathbf{R}_1) = V(\mathbf{r}_i)$$

for any lattic vector, $\mathbf{R}_1$.

Equation 2 would be extremely simple to solve were it not for the electron-electron interaction term, which prevents a separation of the partial differential equation. The approach generally used to circumvent this difficulty is the single-particle approximation,[25] in which each electron is considered as moving in the average potential due to all the other electrons. With this assumption, electronic interaction term just adds a contribution to the potential energy, $V(\mathbf{r}_i)$. We pay a price for this approximation—the effective potential energy now depends on the electronic eigenfunction, which can be determined only by solving the problem. Thus the differential equations for the eigenfunctions must be solved self-consistently rather than directly. In practice, this means that we have to assume a set of eigenfunctions and specify their occupation, and then use these in the single-particle Schrodinger equation to obtain a revised set, and hope that the entire process not only converges, but converges to the true ground state of the system. Furthermore, the approximation neglects completely the possibility that two electrons can correlate their motion in such a way as to keep away from each other spatially, and thus lower their average repulsive energy. On the other hand, we gain an approach that enables us to deal with states of individual electrons, which may be placed in any single-particle state already unoccupied (in accordance with Pauli Exclusion Principle), and which may be treated as mutually independent particles.

In the single-particle approximation, the eigenfunctions of Equation 2 are written

$$\Phi_n(\mathbf{r}_i) = \det |\phi_i(\mathbf{r}_j)| \qquad (3)$$

where $\Phi_i(\mathbf{r}_j)$ is a function depending only on the position of the jth electron. The determinant, (3), guarantees that the wave function changes sign upon the interchange of any two electrons, a fundamental requirement for particles of half-integral spin.

Up until now, everything has been completely general and independent of whether we are dealing with a crystalline or amorphous solid. For a perfectly periodic crystal, we note that the effective single-particle potential is itself periodic, and if we impose periodic boundary conditions[19] to eliminate the disorder due to the presence of surfaces, the eigenstates must be Bloch states, of the form[26]

$$\phi_{n,\mathbf{k}}(\mathbf{r}_i + \mathbf{R}_1) = e^{i\mathbf{k}\cdot\mathbf{R}_1}\, \phi_{n,\mathbf{k}}(\mathbf{r}_i) \qquad (4)$$

where the states are labeled by $\mathbf{k}$, a vector that characterizes the translational behavior of the eigenfuction, and n, a positive integer called the band index. It can easily be shown[27] that $\mathbf{k}$ must be real and restricted to a given set of values called the first Brillouin zone. An important consequence of the reality of $\mathbf{k}$ is

$$|\phi_{n,\mathbf{k}}(\mathbf{r}_i + \mathbf{R}_1)|^2 = |\phi_{n,\mathbf{k}}(\mathbf{r}_i)|^2$$

for any $\mathbf{R}_l$ Thus, any electron is equally likely to be in any primitive cell of the crystal; i.e. all states are delocalized. A second consequence is that the single-particle eigenfunctions have what might be called perfect phase coherence.[14] If the phase of $\Phi_{n,k}$ $(\mathbf{r}_i)$ is taken to be $\theta_{n,k}$ (0) at $\mathbf{r}_i$=0, then the phase at $\mathbf{r}_i$and $\mathbf{R}_l$ is determined from Equation 4 to be

$$\theta_{n,\underset{\sim}{k}}(\underset{\sim}{R}_l) = \theta_{n,\underset{\sim}{k}}(0) + \underset{\sim}{k} \cdot \underset{\sim}{R}_l$$

A qualitative understanding of energy-band theory and the single-particle approximation can be obtained by considering the tight-binding approximation, in which a crystal is built up from the atomic limit—a periodic array of, say, N atoms with infinite nearest-neighbor separation. In this limit, the energy eigenvalues are known precisely, since they are the atomic energy levels. All the electronic energy levels are the same for each of the N atoms; thus any s atomic level is 2N-fold degenerate, any p level is 6N-fold degenerate, etc. As we reduce the lattice parameter sufficiently so that an electron on one atom begins to feel the presence of the nearest-neighboring atoms, these enormous degeneracies are lifted. As long as the interactions between atoms are small, the levels will not move very much in energy, however, so we can say that the atomic levels have spread into a narrow energy band. The importance of this argument is that it does not depend on periodicity in any way. Thus the concept of alternating regions of allowed and forbidden single-electron energies is independent of crystallinity. The delocalization of the electronic states, however, can be established only for periodic systems.

B. Nearly-Perfect Crystals

The introduction of defects of impurities into a perfect crystal destroys the periodicity of the potential energy, and invalidates the applicability of Bloch's theorem. However, in crystals in which there are only small densities of imperfections, it can be assumed that the nonperiodic part of the potential is sufficiently small to be treated as a perturbation. In such cases, the Bloch states are approximate eigenstates and the effect of the imperfections is to introduce a scattering between states of different **k**. We can define parameters such as the lifetime, τ the average time which an electron remains in a Bloch state, or alternatively, the mean free path, Λ, the average distance traversed by an electron in a given state before it is scattered into another Bloch state. The exact eigenstates, of course, are no longer Bloch states, but are linear combinations of the form[14]

$$\phi(\underset{\sim}{r}) = \sum_{n,\underset{\sim}{k}} A_n(\underset{\sim}{k})\ \phi_{n,\underset{\sim}{k}}(\underset{\sim}{r}) \qquad (5)$$

which mixes in states of different **k**. We should expect that the amplitudes, A_n (**k**), are large only over a range of **k** given by Λ^{-1}.

The introduction of scattering also results in a finite mobility. The mobility can be written

$$\mu = \frac{e\tau}{m^*} \qquad (6)$$

where e is the charge and m* is the effective mass of electrons in the solid. Since $\Lambda=v\tau$, where v is the average electronic velocity, mobility is proportional to mean free path. It was first pointed out by Ioffe and Regel[28] that for mobilities less than the order of 1 cm^2/V-sec, the mean free path is less than the interatomic separation. For such cases, the concept of a perturbation theory based on a basis of Bloch states is of doubtful validity. (However, it should be noted that the minimum value of the mobility given above decreases with increasing m*, so that mobilities of, say, 0.1 cm^2/V-sec can be appropriate for narrow-band materials.)

The perturbing potential also affects the energies of the single-particle states, although for weak perturbations we do not expect much change in the density of states. A net attractive potential tends to lower the energies of the unperturbed states, while a net repulsive potential brings about increases in the energy levels. For example, an impurity that is more electronegative than the atom for which it substitutes is likely to provide an excess localized attraction as compared to the perfect crystal. This could cause one or more states to split off from the bottom of a band into the region of forbidden energies. In a semiconductor, a normally filled (valence) band is separated from an empty (conduction) band by an energy gap, E_g. In a perfect crystal conduction occurs by means of thermal excitation of electrons and holes across this energy gap, and the conductivity is given by

$$\sigma = \sigma_o\ e^{-Eg/2\underset{\sim}{k}T} \qquad (7)$$

A relatively electronegative impurity can lead to a conduction-band state being lowered a given energy, E_d, into the gap. This state is then a bound state that is localized around the impurity. If the impurity contains an extra electron as well, the electron, at equilibrium, must occupy the lowest unfilled single-particle state. Since all valence-band states are filled, the extra electron will occupy the state which has been split off from the conduction band. Although it is then a localized electron and thus cannot contribute to the conductivity in the absence of further excitation, it lies much nearer the itinerant conduction band states than any of the electrons in a perfect crystal of the semiconductor. Because of this, at finite temperatures there will be an additional contribution to the conductivity, (7), equal to approximately

$$\sigma_d = N_d \, e^{-E_d/2kT} \quad . \tag{8}$$

where N_d is the concentration (number per unit volume) of these impurity atoms, which are known as donors. Only electrons, as opposed to holes, participate in this type of conduction, which is called n-type.

On the other hand, a vacancy, for example, is ordinarily more attractive than a neutral atom, and should contribute a repulsive perturbation. This repulsion can be expected to raise states from the valence band into the gap. Furthermore, a vacancy represents a decrease in the total number of electrons, as compared to the perfect crystal. Thus these states just above the valence band are initially empty. Excitation of electrons from itinerant valence band states into these low-lying empty states provides a contribution to conductivity analogous to that of (8), except that it is hole, or p-type, conduction. Any imperfection that can bring about p-type conduction is called an acceptor.

Much more precise theories of donors and acceptors in semiconductors exist,[29,30] but it is important to note that (1) the states in the gap represent bound states; (2) consequently they are localized around the imperfection; and (3) the remaining states in the valence and conduction band are, in general, still delocalized, unless the perturbing potential is extremely large.

C. Localized States and Mott Transitions

The energy-band theory of solids, discussed in sub-sections A and B, has been so successful in understanding the properties of nearly all periodic crystals that it is often overlooked that two important assumptions have been made, and these need not always be valid. Firstly, the adiabatic approximation was necessary in order to deal with the electronic problem independent of motion of the ion cores. The breakdown of this approximation will be discussed in sub-section D. Secondly, the assumption of eigenfunctions that are determinants of products of single-particle functions, of the form in Equation 3, completely neglects electronic correlations, as was pointed out in sub-section A. The potential importance of these correlations in perfect crystals was first discussed by Mott,[31-33] who demonstrated a major weakness of the single-particle approximation for the case of extremely narrow, partially filled bands. Mott considered the situation in which a monovalent material, such as Na, was artificially built up from the atomic limit, by bringing together a periodic array of atoms from infinite separation to their proper equilibrium spacing. Let us consider the electronic problem of the case where the interatomic separation is, say, one foot. Since the atomic wave function of the 3s electron of Na decays exponentially with distance from the nucleus, its amplitude is rather small, six inches from the nucleus. However, an exponential is never exactly zero, and if we proceed to apply the single-particle approximation to calculate the band structure, we should find that an extremely narrow, but finite, energy band exists for the 3s electrons. But no matter what the band width, we know exactly how many states exist in the band, two per atom in the material, one spin-up state and one spin-down state. Since only one 3s electron per Na atom is available to be placed in the band, the 3s band is exactly half-filled. Blind application of the band theory of conductivity would lead us to predict that such a hypothetical material would be a metal, an absurd result. It is clear that this "solid" is just a collection of isolated atoms, and in order to transport a single electron through the material, that electron must be removed from one atom and placed on a distant atom. Such a transfer could be accomplished only by supplying an amount of energy equal to the difference between the ionization potential and the electron affinity of the atom. For Na, this difference is about 5 eV, and energy much too large to result in the production of a significant concentration of carriers by means of thermal excitation.

The reason why band theory fails in this case is that the single-particle approximation treats each electron as if it were moving in the average potential due to all the other electrons, assuming this quantity is fixed in time. Thus, as the electron under consideration propagates from one ion core to the next, there is no way of telling whether or not there is another outer electron of opposite spin already present in the ion core.[33] The potential energy of the ion cores, V(**r**), attracts a spin-up electron and a spin-down electron equally, and in the narrow bandwidth limit the states available to electrons of either spin are located at the same energy. However, when two electrons are simultaneously present on the same ion core, we know that a large Coulomb repulsion exists between them. In fact, this repulsion, usually called U, is just the difference between ionization potential and electron affinity discussed previously. The single-particle approach to a widely-separated, periodic array of Na atoms leads to a half-filled, essentially zero-width band. The neglect of correlations results in, on the average, half of the Na atoms containing exactly one 3s electron; one quarter of the atoms have no 3s electrons, and one quarter have two 3s electrons. If the negative ionization potential of the atom is I, the total energy of the ground state (and all other states in which the outer electrons remain in 3s orbitals) in the single-particle approximation is then N(I+U/4), where N is the number of atoms, and the zero of energy is taken to be the energy of an unoccupied set of ion cores. The band structure is given by a zero-width band of 2N states at energy I+U/2.[34]

This example makes it clear just how bad the single-particle approximation can be, since, in this case, we know the exact many-particle energies. The ground state is one in which there is precisely one outer electron per atom, and its energy is thus NI. The first excited state (ignoring any collective excitations) is one in which an outer electron is removed from its ion core and placed on another that already contains an electron of opposite spin; its energy is NI+U. The possible total energies can be written as NI+nU, where n is an integer between 0 and N/2. Thus, the single-particle approximation leads to two serious errors. The ground state energy is overestimated by more than leV per atom, and the ground state is erroneously predicted to be metallic, instead of insulating with a 5 eV energy gap.

It is clear that, in principle, a material can be insulating due to electronic correlations, even when the single-particle approximation predicts metallic behavior. Such a material can be called a Mott insulator. There is now very strong evidence that many transition-metal and rare-earth compounds, such as CoO and Eu_2O_3, are Mott insulators.[23,35] The importance of this is that the best single-particle eigenstates, even for perfectly periodic crystals, are not always Bloch states. For Mott insulators, a more accurate basis set with which to analyze the electronic problem is the set of localized atomic orbitals.[36] Consequently, in these materials, the existence of long-range positional order is completely irrelevant. In recent years, several quantitative calculations have been performed,[37-42] all indicating that correlations can indeed lead to localization in periodic systems, just as Mott originally postulated.

Mott[32,43] also asked the question of how the transition from localized to Bloch states comes about in a periodic crystal, as the interatomic separation is continuously decreased. Physically, the metallic state of Na must result from an effective screening of the intra-atomic Coulomb repulsion, U, by all the other electrons in the crystal. It is clear that this screening will increase sharply as the interatomic spacing decreases, and eventually an insulator-metal transition occurs. Mott presented a simple argument in favor of a sharp transition to the metallic state.[32] In a Mott insulator, a single free electron and a single free hole attract each other with a Coulomb interaction,

$$V_c(r) = - e^2/\varepsilon_o r \quad , \tag{9}$$

where ϵ_o is the static dielectric constant of the insulator and r is the electron-hole separation. This interaction has bound-state solutions, and the lowest energy state will be a hydrogenic bound state, called an exciton, in which neither electron nor hole participates in conduction. On the other hand, if many free carriers are present, the Coulomb attraction between electrons and holes, Equation 9, becomes screened, and must be replaced by[44]

$$V_s(r) = - (e^2/\varepsilon_o r)\ e^{-\alpha r} \tag{10}$$

where a is a screening constant, which increases monotonically with the carrier concentration. For sufficiently large a, the potential energy, Equation 10, produces no bound-state solutions, and a sudden transition occurs from no carriers participating in conduction to a large number. This insulator-metal transition is now known as a Mott transition.

D. Electron-Phonon Interactions and Polarons

The adiabatic approximation enables us to separate the electronic problem from the lattice problem. Up to this point, we have been discussing only the former. The lattice problem, in principle, uses the electronic energy levels, calculated as a function of the fixed position of the ion cores, as the potential energy for ionic vibrations. In practice, an empirical approach is always used, in which it is assumed that the ions undergo only small oscillations around their equilibrium positions (as determined from diffraction experiments), the potential energy for such vibrations being determined by the observed force constants of the material. The lattice vibrational quanta are called phonons, in analogy with the particle interpretation of electromagnetic waves, (photons). It is found that there always exist low-frequency long-wavelength oscillations that approach rigid displacements of the entire crystal. These are called acoustic phonons, in analogy with the long-wavelength oscillations (sound waves) of an elastic continuum. If there is more than one type of atom in the material, or if there are inequivalent atoms of the same type, there are also high-frequency, long-wavelength oscillations, in which the inequivalent atoms vibrate against each other. The quanta of these vibrations are called optical phonons.

In an ionic crystal, at least two types of atoms are present and optical phonons must exist. During a longitudinal optical lattice vibration, positive and negative charges move in opposite directions, with the amplitudes of the motion varying with distance along the direction of propagation. This thus sets up alternating regions of positive and negative charge density in excess of the equilibrium values. Consider an itinerant electron with most of its amplitude near a particular point in the primitive cell of an ionic crystal. The electron, being a negative charge, can lower its energy by inducing the longitudinal optical lattice distortion that brings extra positive charge near the regions where its amplitude is large and moves negative charge where its amplitude is small. It is clear that in ionic crystals electrons interact strongly with longitudinal optical phonons. But the adiabatic theorem assumes that such interactions are negligible, since it separates the electronic problem from the phonon problem. The existence of strong electron-phonon coupling in ionic crystals represents another breakdown of ordinary energy-band theory of solids, even in perfect crystals. This problem is handled by considering not single-electron states, but rather states of the electron together with its induced longitudinal optical lattice distortion, a quasiparticle known as a polaron. The polaron has lower energy than the corresponding electron, but it has a larger effective mass, since it must carry its associated lattice deformation with it as it moves through the crystal.

Polaron theory has been derived in various limits, depending on the strength of the electron-phonon interaction, the extent of the lattice distortion, and the bandwidth of the electrons. In the limit in which the associated lattice deformation extends over a large number of lattice parameters, the lattice can be replaced by a continuum, as in the Debye model for specific heats or in the theory of dielectric constants. It can be shown[45] that the extent of the lattice distortion induced by an electron of effective mass, m*, is

$$r_o = \left(\frac{\hbar}{2m^* \omega_o}\right)^{\frac{1}{2}} \tag{11}$$

where ω_o is an average longitudinal optical phonon frequency; r_o is called the radius of the polaron. If r_o is much larger than the interatomic spacing the polaron is called large. In large-polaron theory,[46,47] the parameter which gives the strength of the coupling between electrons and longitudinal optical phonons is defined as

$$\alpha = \frac{1}{2}\left(\frac{1}{\varepsilon_\infty} - \frac{1}{\varepsilon_o}\right)\frac{e^2}{r_o}\frac{1}{\hbar\omega_o} \tag{12}$$

where ϵ_o is the static dielectric constant and ϵ_∞ is the high-frequency dielectric constant. For $a < 6$, the coupling is called weak, and the single-electron energy levels are modified only slightly. The electronic energy is decreased by $a\,\hbar\,\omega_o$, and the band effective mass is increased by a factor of $(1 + a/6)$. For $a > 6$, strong-coupling theory[48] applies, and energy-band theory must be modified significantly. The reduction in electronic energy can be estimated as $0.1\ a^2$, and the effective mass increases by a factor of $0.02\ a^4$. Thus, for very large a, the polaron appears to be so massive that it hardly moves in an applied electric field. This results in a serious suppression of the electrical conductivity of the crystal. But note that the large a can be obtained only by a small optical-phonon frequency, ω_o, if large-polaron theory is to remain applicable. As can be seen from Equations 11 and 12, a large band effective mass also leads to a large α but implies a small polaronic radius, r_o, inconsistent with the continuum approximation.

In the opposite limit, in which r_o, is small compared to the interatomic separation, a, the continuum approximation cannot be used. If, in addition, the energy

$$E_p \equiv \frac{1}{\pi^2} \frac{r_o}{a} \alpha\hbar\omega_o \tag{13}$$

is large compared with the electronic bandwidth, Δ, the quasiparticle is called a small polaron.[49,50] Since a small Δ implies a large m*, and thus a small r_o, small-polaron theory applies in very narrow-band materials. The quantity E_p defined in Equation 13 is the approximate reduction of electronic energy due to small-polaron formation.

As opposed to the situation in which the polarons are large, small-polaron formation results in qualitative modification of the electronic transport properties. Holstein[50] found that at low temperatures, small-polaron states overlap sufficiently to form a polaron band in which ordinary conduction takes place. However, the width of this band decreases exponentially as the temperature is raised, and above a critical temperature, T_t, the bandwidth is less than the uncertainty in energy due to the finite polaron lifetime. Above T_t, conductivity in the polaron band is negligible, and the small polaron can be considered to be localized. As we noted in sub-section C, extremely narrow-bandwidth materials are Mott insulators. Thus, in Mott insulators, the carriers almost undoubtedly form small polarons, and are themselves localized. Once again, long-range periodicity is irrevelant for small polarons above T_t.

Localization of potential free carriers in an ionic crystal was first discussed by Landau,[51] who noted that an excess electron moving in the vicinity of a particular lattice point will attract the positive ions and repel the negative ions, thus inducing a local lattice deformation. The deformation yields a potential well with which the electron can form a bound state. It is this bound state which is the localized small polaron. Once the polaron is localized, it can still contribute to conductivity, by means of hopping between equivalent sites. Because the local lattice deformation must move through the crystal, the hopping process requires an activation energy above that required to create the free carrier. In other words, if we write the conductivity as

$$\sigma = ne\mu, \tag{14}$$

where n is the free carrier concentration and μ is the mobility, then when hopping dominates the conductivity

$$\mu = \mu_o e^{-E_a/kT} \tag{15}$$

where E_a is the energy necessary for the polaron to hop between two lattice sites. Holstein[50] showed $E_a = E_p/2$, where E_p is given by Equation 13.

For a wide range of the electron-phonon coupling constant, α, and the bandwidth, Δ, Holstein[50] estimated the transition temperature, T_t, as approximately half the Debye temperature of the crystal. However, for very narrow bandwidths, it can be shown[52] that T_t becomes quite small, and hopping conduction predominates essentially throughout the entire temperature range.

In principle, it should be simple to distinguish between ordinary band semiconduction and hopping conduction. When hopping predominates, the mobility must exponentially increase with temperature, as indicated by Equation 15. The temperature dependence of the mobility of an electron or polaron moving in an energy band depends on the predominant scattering mechanism, but could be an algebraic increase (impurity scattering), an algebraic decrease (acoustic-phonon scattering), or an exponential decrease (optical-phonon scattering). The contribution to conductivity from mobility can be separated from that carrier concentration by combining conductivity measurements with results of Hall effect or thermoelectric power experiments. If only one type of carrier contributes to electrical transport in a band, then the Hall mobility, defined as

$$\mu_H \equiv c\,|R_H|\,\sigma \tag{16}$$

where R_H is the Hall constant and c is the velocity of light, should be proportional to the conductivity mobility, μ. For small-polaron hopping, however, the Hall mobility should increase exponentially with increasing temperature, with an activation energy of $E_a/3$.[53] Alternatively, if the conduction is dominated by free electrons, the thermoelectric power, S, can be expressed as

$$S = -\frac{k}{e}\left[\frac{E_c - E_f}{kT} + a\right] \quad , \qquad (17)$$

where E_c is the conduction band edge, E_f is the Fermi energy, and a is a constant (usually between 2 and 4) that takes into account the fact that the moving electrons have an average energy somewhat greater than that of the bottom of the conduction band. If two types of carriers contribute to the conduction, the thermoelectric power is the difference between two terms,

$$S = \frac{k}{e}\left[\frac{\sigma_h}{\sigma}\left(\frac{E_f - E_v}{kT} + a_h\right) - \frac{\sigma_e}{\sigma}\left(\frac{E_c - E_f}{kT} + a_e\right)\right]$$

where σ_h and σ_e are the contributions to conductivity from holes and electrons, respectively. For the case of the hopping among localized states near the Fermi energy, Mott[6] has suggested the metallic expression

$$S = \frac{\pi^2 kT}{3e}\left[\frac{\partial \ln \sigma (E)}{\partial E}\right]_{E=E_F}$$

Measurements of the temperature dependence of S should be able to distinguish between band-like and hopping conduction. However, this determination has proved to be extremely difficult in practice, primarily because of spurious contributions to the Seebeck coefficient.

III. AMORPHOUS – SEMICONDUCTOR THEORY

Although a rigorous, quantitative theory of amorphous semiconductors does not exist at present and there is little hope of obtaining first-principles solutions of the Schrodinger equation for a disordered system in the near future, nevertheless, great strides have been made in the formulation of a semi-quantitative approach, which enables us to analyze the experimental results in a consistent manner. As indicated in section I, we are interested in obtaining the one-electron density of states g(E) and the average mobility $\mu(E)$ for amorphous systems. Since $\mu(E)$ depends on g(E), an evaluation of the density of states is of primary importance.

The key to estimating g(E) is the short-range order of amorphous systems. From the viewpoint of the tight-binding approximation[54] of ordinary band theory, the electronic energy levels of a solid depend primarily on the nature of the atom from which the electron originates and on the nature and positions of its nearest neighbors. As has been discussed, these may be essentially the same in corresponding crystalline and amorphous solids. The effects of slight variations in nearest-neighbor orientations and large changes in the positions of the second and farther neighbors in the amorphous material can then be introduced as a perturbation on the crystalline density of states, which can be calculated quantitatively. Consider the simplest possible density of states for a single, non-degenerate band of a crystal, shown in Figure 3a. The sharp structure, both at the band edges and internally, represents the van Hove singularities,[55] which are a necessary consequence of three-dimensional periodicity. If we introduce structural disorder as a non-k-conserving perturbing potential, we should expect all the states to shift somewhat in energy, some up and some down. The extent of these shifts clearly depends on the amount of disorder present, but in any event the perturbation removes the van Hove singularities. In particular, the band edges will no longer be sharp, and states will extend beyond the original band extrema. The density-of-states of a band in an amorphous solid is shown in Figure 3b.

Those parts of the band of Figure 3b that have energies below the minimum and above the maximum energies of the corresponding crystalline band, Figure 3a, are called band tails. Band tails due to disorder are not a recent discovery. They appear to have been first studied by Lifshitz,[56-63] and later by Fröhlich[64] and many others,[65-79] primarily with regard to impurity-induced disorder in crystalline solids. With a Gaussian approximation for the perturbing potential,[80] the density-of-states deep in the band tail is

$$g(E) \sim |E|^{-3/2} \exp(-E^2/2\langle U^2\rangle) \quad , \qquad (18)$$

where $\langle U^2\rangle$ is the mean-square perturbation. Thus, although the fall-off is very rapid in this approximation, the extent of the band tail is proportional to the root-mean-square deviations from periodicity. This is an intuitively obvious but physically important result. From it, we can

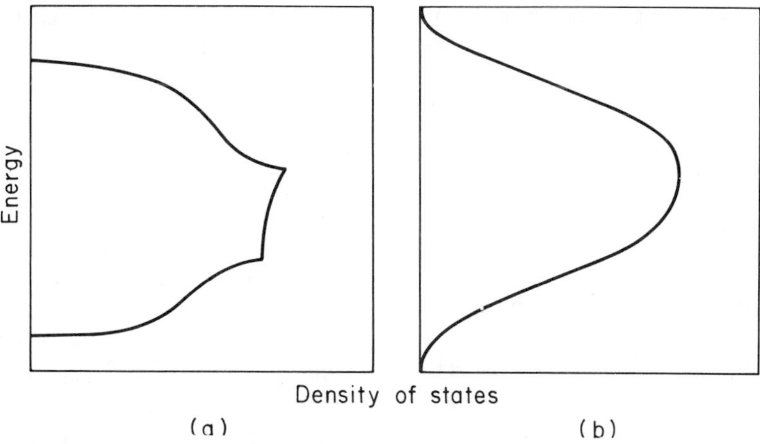

FIGURE 3. (a) Density of electronic states as a function of energy for a single band of a crystalline solid. The sharp behavior at the band edges and in the interior represents the effects of Van Hove singularities. (b) Density of electronic states as a function of energy for a single band of an amorphous solid. All Van Hove singularities have disappeared.

conclude that the band tailing should not be very extensive in, for example, amorphous Ge and Si, but, on the other hand, we should expect enormous tailing in the multicomponent chalcogenide glasses.

The result for g(E) given in Equation 18 is, of course, entirely due to the Gaussian form of the perturbing potential. Although this is the most common approximation used in calculating band tails of disordered systems, it is not particularly physical. Another class of approximation is based on modeling an amorphous solid as a disordered alloy. This is essentially a substitution of compositional for positional disorder, and the validity of such an approach is not obvious. The great simplification brought about by the alloy approximation is that the periodic lattice is maintained. Disorder can then be introduced either by randomly arranging two different types of atoms on all of the sites, thus contributing atomic potentials V_A $(\mathbf{r}\text{-}\mathbf{R}_i)$ and V_B $(\mathbf{r}\text{-}\mathbf{R}_j)$, or by considering a fluctuating atomic potential, V_o $(\mathbf{r}\text{-}\mathbf{R}_i)$ + δV_i $(\mathbf{r}\text{-}\mathbf{R}_i)$, where δV_i varies from site to site. The former is appropriate for the binary alloy problem, the latter is a more realistic approximation for an amorphous solid. However, in an actual amorphous semiconductor, the atomic positions R_i, do not form a periodic system, and the situation is still more complicated.

An important technique for quantitatively estimating g(E) for the binary alloy problem is the coherent potential approximation (CPA), first introduced by Lax,[81] and developed in some detail by Soven.[82] This approximation replaces the real potential by a periodic potential plus a perturbation which is chosen so that the average scattering is the same as in the actual solid. This can be done provided the mean free path is large compared to the DeBroiglie wavelength of the propagating carrier. This condition, unfortunately, is never satisfied in amorphous semiconductors. A further disadvantage is that the method is too insensitive to exhibit any band-tailing effects, due to the fact that the approximation begins to break down near the band edges. Butler[83] has shown that a generalization which treats interference of waves scattered from adjoining sites should be appropriate in the opposite limit, in which the mean free path is of the order of the interatomic spacing. This follows from the observation that in such a situation, physical quantities at two points separated by a large number of interatomic distances are essentially entirely uncorrelated.[84] In fact, it can be shown that an approximation for the density of states of a disordered system can be obtained by averaging over possible atomic configurations in a single unit cell to accuracy $e^{-\Lambda/a}$, where Λ is the mean free path and a is an average interatomic spacing. However, since the band tail arises solely from fluctuations in local order or composition, and the generalized CPA averages over these fluctuations, sharp band edges still result.

Another generalization has been suggested by Freed and Cohen,[85] who investigated a cluster theory in which the CPA is the first-order approximation. By averaging only over atoms outside variable-sized clusters, Freed and Cohen defined a hierarchy of Green's functions and showed that the equation of motion for the n^{TH} nearest-neighbor Green's function is coupled to the $(n+1)^{ST}$ such function. Few quantitative results have as yet been extracted from this model, but successive truncations of the Green's function equations should give a well-defined series of approximations which converge to the exact answer.

Another simple type of approximation for g(E) was introduced by Gubanov,[74] who suggested a coordinate transformation which continuously undistorts the amorphous system as a small parameter is varied, so that the system is perfectly periodic in the new coordinates. Although such a procedure leads to a band tail in the density of states, it is unphysical in the sense that both the metastability and the short-range order of amorphous solids are neglected.

We have discussed qualitatively in section I how we can define an average mobility funcion μ (E), which can be expected to be relatively large in regions where g(E) is large but should be essentially zero in regions where g(E) is small, i.e. in the band tails.[86] The behavior of μ (E) in regions of intermediate density of states is still a matter for speculation. One possibility is that μ (E) gradually decreases from a finite value to zero over a wide range of energy, much as g(E) itself behaves. Another possibility is that μ (E) falls to zero sharply, but continuously, over a narrow range of energy. Finally, μ (E) could discontinuously vanish at a single critical energy. All of these possibilities are shown in Figure 4. The real behavior of μ (E) at intermediate densities is a very difficult problem, which can be resolved only by detailed calculations.

The first attempt to quantitatively discuss μ(E) for a disordered system was by Anderson[87] in a paper that is now considered the foundation of amorphous-semiconductor theory. Anderson analyzed the problem of a three-dimensional periodic array of square wells of random depths spread over a range of energy, V_o, as shown in Figure 5. Let Δ be the spread in energy that would exist if the ground states of an array of identical wells at an average energy E_o were allowed to overlap and form a band. For the disordered system, the bandwidth increases to $\Delta + V_o$. Anderson showed that above a critical ratio of V_o to Δ, an electron placed in one of the wells never completely diffuse away. The calculation is extremely complex, and is based on a Brillouin-Wigner perturbation expansion for the electronic energy, with an array of independent wells as the unperturbed system and the overlap integral as the perturbation. The convergence of the perturbation expansion was taken as the condition for lack of diffusion, and thus for localization. A major computational problem is that the existence of many different states with energies very near each other can lead to an unphysical divergence of the perturbation series even when the states are all spatially well separated and thus strongly localized; multiple scattering from sites of essentially the same energy leads to products of large energy denominators which dominate the expansion. To avoid this difficulty, Anderson applied a scattering formulism suggested by Watson,[88] which eliminates the divergences by self-consistently recalculating the energy levels and thus preventing multiple scattering from sites at nearly the same

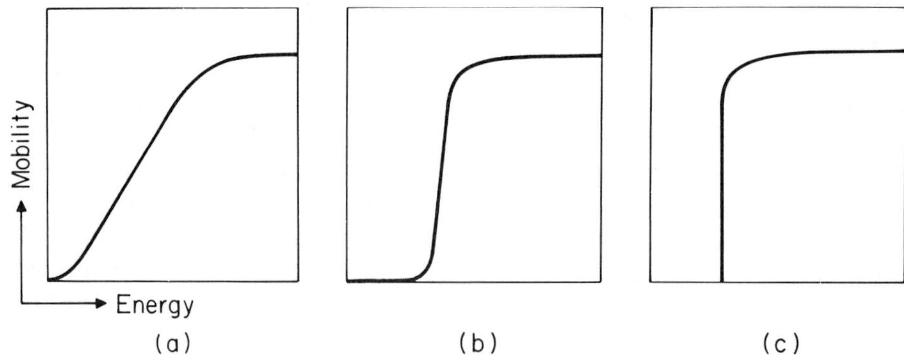

FIGURE 4. Three models for the increase of mobility as the energy increases from the band tail to the dense portion of the band.

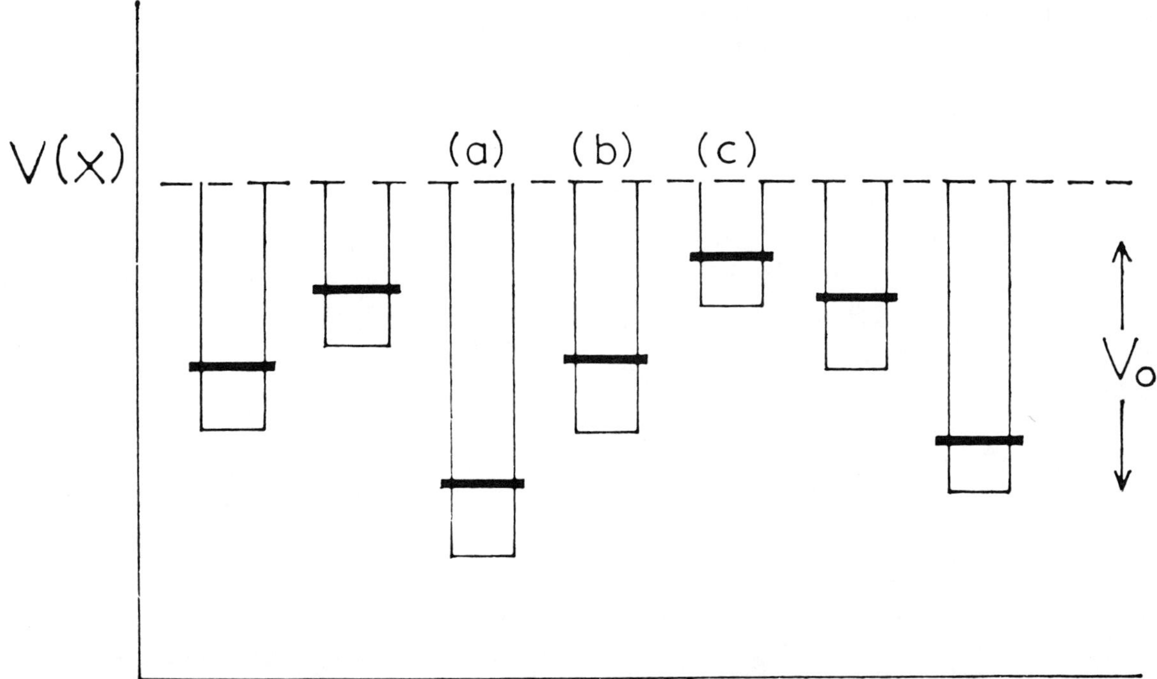

FIGURE 5. Potential wells arising from the ion cores in the Anderson model. The potential energy of an electron is plotted against distance. The heavy horizontal lines represent the energy of the corresponding state (Mott[98]).

energy. With this formulism, Anderson was able to estimate an upper bound for the critical ratio of V_o to Δ, for a simple cubic lattice, to be

$$\left(\frac{V_o}{\Delta}\right)_c < 4.3 \quad . \tag{19}$$

He further speculated that a more exact procedure that takes into account correlations in the disordering potential would lead to a critical ratio of

$$\left(\frac{V_o}{\Delta}\right) \approx 2.8 \quad . \tag{20}$$

The critical ratio varies somewhat with the lattice under consideration, but the exact value is not as important as its very existence. By demonstrating that a critical value exists, Anderson showed that disorder can indeed lead to localization.

Anderson's model is computationally extremely difficult, but it has recently been criticized on more fundamental grounds.[89-93] Bonch-Bruevich[89] suggested that the mobility at T=O can never vanish unless the density of states at the Fermi energy vanishes, a theorem in direct conflict with that of Anderson. Bonch-Bruevich's result also appears to be in conflict with experiment, since conduction at low temperatures in partly compensated semiconductors has been shown to be by thermally activated hopping, and thus the states are localized. Lloyd[91] presented arguments that an exact solution to Anderson's problem for a Lorentzian distribution of random potentials gives no localization, essentially because the density of states g(E) has no singularity. However, the arguments we have presented in section I indicate that no such singularity is necessary to obtain sharp behavior in $\mu(E)$. Furthermore, the quantitative calculations referred to in section II with regard to a Mott transition show that a sharp mobility jump can occur at a critical g(E). Finally, Brouers[92] and Ziman[90] have attempted to point out errors in Anderson's mathematical procedure. These papers, however, average over all possible atomic configurations at an early point in the calculation, much as in the CPA.[81] As Anderson[94,95] and Thouless[96] have pointed out, such an averaging eliminates the possibility of obtaining localized states by guaranteeing divergence of the perturbation expansion. In other words, localization depends on the fact that the environment of any site is unique, and since any averaging procedure destroys the uniqueness, no conclusions about the mobility should be inferred from such a technique.

Anderson[87] considered only states near the center of the band. When they become localized, all other states in the band are also localized.

Because of the short-range order always present in amorphous semiconductors, it is highly unlikely that the critical ratio of disorder to band width is ever exceded in the ordinarily wide outer bands of a solid. Thus, in most cases, states near the center of the band are delocalized. As we previously discussed, it is physically evident that states far out in the band tail must be localized, since the density of such states is very small. These states exist solely because of a particularly strong attractive or repulsive potential fluctuation resulting from extreme structural aberration, and clearly no overlap exists between the wave functions localized around these far-removed regions. Mott[6,77,97-103] was the first to emphasize that these considerations necessitate a mobility transition of one of the types shown in Figure 4. Mott[97] suggested, primarily on the basis of the critical point obtained in the Anderson model, that the sharp transition of Figure 4c was the most likely. Edwards,[104] in a simple calculation, found a similar result, although Cohen[105] was led by the results of a one-dimensional model[106,107] to conclude that the less sharp transition of Figure 4b was more likely. In either event, we can identify a critical energy which separates localized from delocalized states. Such an energy has been called a *mobility edge,*[108] and it plays the same role in disordered systems that the band edge plays in periodic solids.

The existence of mobility edges in amorphous solids now has very strong theoretical support and, in addition, provides a straightforward interpretation of a large variety of experimental results, as we shall see in sections IV, V, and VII. Recently Economou, Cohen, and co-workers[109-111] have demonstrated the existence of sharp mobility edges in the Anderson problem for several special cases. By introducing a convergence function, L(E), such that L<1 for energies at which μ(E) essentially vanishes and L>1 for energies at which the states are delocalized, they were able to solve for the mobility edges by setting L(E)=1. This was carried out for the case of a Lorentzian distribution of atomic energy levels. The solution to the problem exhibited the entire transformation from an ordinary band in the ordered case, through the formation of band tails with mobility edges, to the Anderson transition at which point all states become localized. The Anderson transition occurs at a critical value of the Lorentzian half-width, Γ, such that

$$\left(\frac{\Gamma}{\Delta}\right)_c \approx 0.5 \quad . \tag{21}$$

For the case of a rectangular distribution of atomic energy levels of width V_o, Economou and Cohen[111] essentially confirmed Anderson's speculation, (20). In addition, Economou et al.[110] used the CPA self-energies calculated for the disordered binary alloy problem to demonstrate that mobility edges and the Anderson transition exist in this special case as well. In binary alloys, introduction of disorder splits off a sub-band from the ordered band, and only the sub-band exhibits localization and an Anderson transition.

All the approaches discussed thus far are based on models in which the atomic sites from a periodic lattice. As we already noted, these approximations are more suited to the disordered alloy problem than to the amorphous-semiconductor problem, in which long-range structural disorder is a necessary feature. A different approach was used by Neustadter and Coopersmith,[112] who considered the problem of randomly arranged identical hard-core atoms which simply scatter electrons with a scattering length small compared to the interatomic separation. Using a solution to the multiple scattering problem obtained by Coopersmith[113] for electrons in helium, Neustadter and Coopersmith calculated the mobility as a function of atomic density at constant temperature and scattering length. When all orders of multiple scattering are taken into account, they found that a critical density exists at which point the mobility sharply drops from about 10^4 cm^2/V-sec to zero. No corresponding density of states was calculated, but the critical atomic density of about 10^{20} cm^{-3} could be associated with the mobility edge for this problem. Eggarter and Cohen[114] treated the same problem by replacing the atomic scatterers by their equivalent potential barriers.[115] The resulting density of states was parabolic far from the edge, and a Gaussian band tail was found. A critical energy, somewhat below the parabolic band edge, was found, below which all states are localized. For a constant atomic density of 10^{20} cm^{-3}, the mobility as a function of energy was calculated, using a semiclassical percolation technique. A linear increase in the mobility from zero was found, reflecting primarily an increase in the percentage of delocalized states with increasing energy above the edge. A fully quantum-mechanical calculation would, of course, forbid the existence of any strictly localized states above the mobility edge, since the would-be localized electron could tunnel into a delocalized state and thus propogate. However, Eggarter and Cohen postulated that these states make only a small contribution to the mobility. This situation is analogous to the problem of a weak attractive perturbing potential in an otherwise periodic crystal. If the perturbation is insufficiently strong to split off a localized (donor) state from the continuum, a delocalized resonant state is produced. Eggarter and Cohen also calculated the mobility as a function of atomic density by the same semiclassical technique, and confirmed the general behavior found by Neustadter and Coopersmith.[112]

An interesting feature of the model of Eggarter and Cohen[114] is that the band tail arises solely from regions in space where the atomic density is anomalously small. Physically, this comes about because in such regions the repulsive potential due to the scatterers is absent, and the electron can consequently attain a lower energy than in the high-density regions. However, the potential barriers then prevent such a low-energy electron from leaving the low-density region, and the state is thus localized. This provides a physical explanation of how the existence of disorder can *lower* the energy of some of the states over what they would be in a perfectly periodic system. This is a point that we shall return to later.

The consequences of band tailing and mobility edges on the electrical conductivity of a semiconductor are straightforward. Figure 6 shows a possible density of states diagram for an amorphous semiconductor, with the valence and conduction band mobility edges labeled E_v and E_c respectively. For a two band solid, the Kubo-Greenwood formula for electrical conductivity[116,117] can be put in the form

$$\sigma = \int_{E'_c}^{E''_c} dE\, g_c(E)\, e\, \mu_c(E)\, \{\exp[(E - E_f)/kT] + 1\}^{-1} + \int_{E'_v}^{E''_v} dE\, g_v(E)\, e\, \mu_v(E)\, \{\exp[(E_f - E)/kT] + 1\}^{-1} \quad (22)$$

where g(E) is the appropriate density of states, μ(E) the average mobility, and E′ and E″ the band minima and maxima, respectively. The Fermi energy is set by the requirement that the correct electronic density is always obtained. Since a semiconductor has a sufficient concentration of electrons to exactly fill every valence-band state while leaving all conduction-band states empty, at T=O we must have

$$E_f = \frac{1}{2}(E''_v + E'_c) \quad . \quad (23)$$

The existence of mobility edges then shows, for $kT < (E_c - E_v)$ that the conductivity takes the form

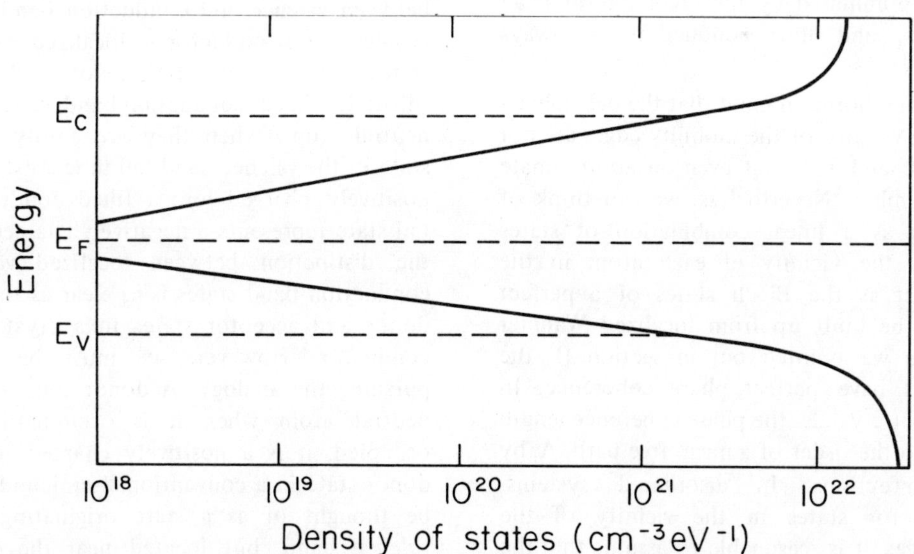

FIGURE 6. Density of states as a function of energy for an amorphous semiconductor. E_v and E_c denote the positions of the valence and conduction band mobility edges, respectively; E_f is the zero temperature Fermi energy.

$$\sigma = \sigma_c e^{-(E_c - E_f)/kT} + \sigma_v e^{-(E_f - E_v)/kT} \quad (24)$$

where σ_c and σ_v can be determined from Equation 22, and E_f is initially determined by Equation 23. An important result is that the Fermi energy at T=O, given by (23), is half way between the valence and conduction *band* edges, not half way between the *mobility* edges. Thus, in general, if the mobilities in the delocalized parts of the valence and conduction bands are not too different, the conductivity will be dominated by either holes or electrons even in the intrinsic range, depending only on whether the conduction or valence band-tail is more extensive. All other things being equal, it is a general quantum-mechanical result that the higher-energy bands are wider than the lower-energy bands. This is simply a consequence of the fact that a faster particle is always scattered less by a fixed potential than a more slowly moving particle, and is an effect already evident in the Kronig-Penney model.[118] In a covalent semiconductor, in which the valence and conduction bands arise from corresponding bonding and antibonding orbitals, all other things may indeed be equal. If this is the case, the conduction band tail is more extensive than the valence band tail; i.e. $(E_c - E_c') > (E_v'' - E_v)$. Then, from Equation 23, we can conclude:

$$(E_c - E_f) > (E_f - E_v) \quad . \quad (25)$$

When (25) is valid, Equation 24 for the conductivity is dominated by the second term at all temperatures, and thus conduction is always p-type.

It should be borne in mind that the delocalized states in the vicinity of the mobility edges are not Bloch states, and **k** is not even an approximate quantum number. Nevertheless, we can think of these states as a linear combination of states localized in the vicinity of each atom in the material, just as the Bloch states of a perfect crystal can be built up from localized Wannier states.[25] As we pointed out in section II, the Bloch states have perfect phase coherence. In nearly perfect crystals, the phase coherence length is reduced to the order of a mean free path, Λ, by the imperfections. In disordered systems, particularly for states in the vicinity of the mobility edges, it is reasonable to assume that the mean free path is the order of the interatomic separation, a, and that the phase coherence essentially disappears entirely. Thus, the delocalized functions can still be written in the form of Equation 5, but the amplitudes of the A_n (**k**) will be large for all **k**, and the phases can be treated as being completely random. Far beyond the mobility edges, it is possible that the eigenstates are approximate Bloch states, and that **k** can be an approximate quantum number, although this is of no importance to equilibrium transport.

Cohen, Fritzsche, and Ovshinsky[108] (CFO) suggested that in cases of extreme disorder, e.g. the chalcogenide glasses, it is possible that the valence and conduction band tails actually overlap in what formerly was the energy gap, as is shown in Figure 7. Such a situation preserves a gap between the valence and conduction band mobility edges, and it is the existence of this mobility gap, $E_c - E_v$, which leads to semiconducting rather than metallic behavior.

There are several unusual features of the CFO model. When conduction and valence band states overlap, it is natural to ask whether we can still differentiate between the two types of states. In principle, since these are localized states, we always can. The reason stems from the fact that a semiconductor ordinarily contains just enough electrons to fill the valence band while leaving states in the conduction band empty. This situation corresponds to saturating all valence requirements, with local neutrality maintaining everywhere.

This focuses on an important distinction between valence and conduction band states in a covalent semiconductor – localized valence-band states represent neutral atoms only when they are filled; localized conduction-band states represent neutral atoms when they are empty. An empty state in the valence band tail thus must represent a positively charged ion; a filled conduction-band tail state represents a negatively charged ion. Thus the distinction between localized valence and conduction band states is as clear as that between donor and acceptor states in a crystalline semiconductor. However, we must be careful in pursuing the analogy. A donor state represents a neutral atom when it is occupied; when unoccupied, it is a positively charged ion. Thus a donor state in a conventional semiconductor must be thought of as a state originating from the *valence* band, but located near the conduction band edge. With the above prescription, we can always identify the parentage of localized states.

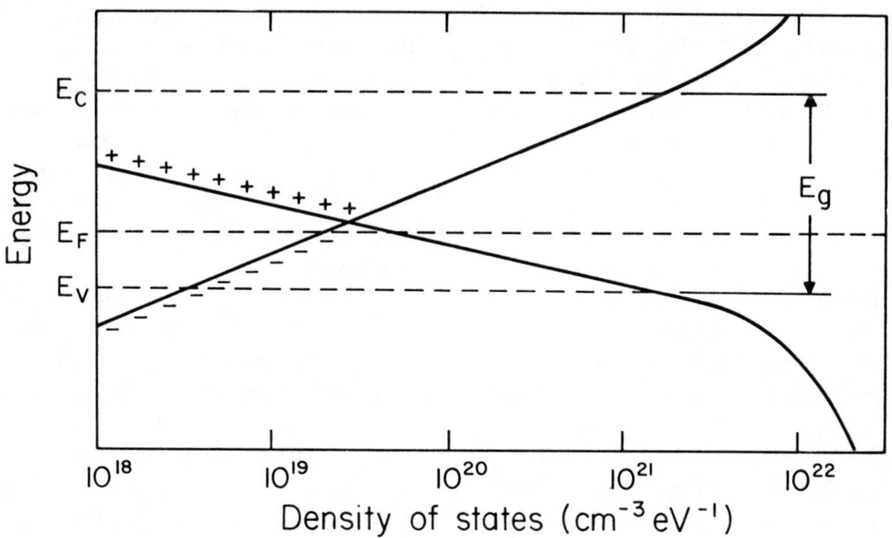

FIGURE 7. CFO model for the density of states of a sufficiently disordered amorphous semiconductor. E_v and E_c are the respective valence and conduction band mobility edges; E_g is the mobility gap and E_f is the Fermi energy. The positions of the positively and negatively charged traps at zero temperature are schematically noted.

Note that the fact that the wave function of the state must be constructed from an admixture of the Bloch states of both valence and conduction bands of the corresponding crystalline solid is not relevant to the question of parentage. As discussed semiquantitatively by Srinivasan and Cohen,[119] all valence and conduction band states are renormalized in the presence of even a single imperfection in a crystal, but the parentage can clearly be determined.

The second unusual feature of the CFO model is that there are states in the valence band, hence ordinarily filled, which have higher energy than ordinarily unfilled states in the conduction band. Clearly this is not an equilibrium situation, and a redistribution of electrons must take place. If we assume that the energy-band structure is independent of occupation, then the highest energy valence-band electrons will simply fall into the lowest energy conduction-band states until the minimum total energy is obtained. The actual ground state is as shown in Figure 7. The filled states in the conduction-band tail represent negatively charged ions; the empty states in the valence-band tail represent positively charged ions. Since the material contains a sufficient number of electrons to fill all valence-band states and leave all conduction-band states unoccupied, there must be equal densities of positively and negatively charged ions after the redistribution, provided all valence requirements of the constituent atoms are locally fulfilled. If dangling bonds exist, perhaps because of the particular method of preparation of the material, some donor or acceptor states will be present, and the concentrations of the two types of charged ions need not be equal. The positive and negative ions act as effective traps for electrons and holes respectively, and also act as scattering centers that could limit the mobility of free carriers. Because of this, it might be expected that mobility increases with increasing energy above the edges.[120,121]

An important assumption of the CFO model is that both the valence and conduction bands tail extensively into the mobility gap. It is relatively straightforward to understand how states in the valence band can increase their energy because of the disorder, thus forming a large band tail. Since the crystalline ground state must have lower total energy than the ground state of an amorphous solid of the same composition, it would be expected that introduction of disorder should increase the energy of the filled states. But equally well, it might be expected that disorder would also raise the energy of the conduction-band states, and thus decrease the likelihood of the band overlap suggested by the CFO model. However, the hard-core scattering model of Eggarter and Cohen[114] suggests that states in the conduction-band tail arise from regions of the material in which the density is anomalously low. An analogous result may indeed be applicable to the covalent amorphous semiconductor as can be seen from Figure 8. The valence and conduction bands of covalent semiconductors arise from bonding and antibonding orbitals of the constituent molecules. Figure 8 shows the energy of the bonding and antibonding orbitals of a hydrogen molecule as a function of interatomic separation. We can infer from this behavior that a deviation in either direction from the equilibrium interatomic spacing contributes to the valence-band tail, but that only local *decreases* in density lead to a conduction-band tail. However, the asymmetry in the energy vs. interatomic separation curve for the bonding orbitals suggests that states deep in the valence band tail primarily arise from local density increases rather than decreases. Thus, if the valence and conduction band tails overlap, the resulting positively and negatively charged ions are spatially separated.

The CFO model is based on the one-electron approximation, which neglects electronic correlations. However, as discussed in section II, electronic interactions are treated in an average manner, and these are extremely important when the charge redistribution takes place. The CFO ground state must be calculated in a self-consistent manner, and the one-electron potential, $V(\mathbf{r}_i)$ in equation (1), will reflect the actual charge density after the redistribution. One of the consequences of the CFO model is that $V(\mathbf{r}_i)$ is a sensitive function of the one-electron occupation numbers, and therefore we must be very cautious in determining the average occupancies of the state as a function of temperature. Ordinary Fermi statistics depend on the relative insensitivity of $V(\mathbf{r}_i)$ to occupation numbers, and this is not the case when localized states are involved.

Fortunately, once again we have past experience with an analogous problem to guide us. The impurity problem in a crystalline semiconductor is instructive in this case, since often (e.g. Au in Ge) several types of localized states representing different valences can exist. Often these are plotted on an ordinary density of states diagram, the donor states being placed below the conduction band and the acceptor states above the valence band by the amount of their ionization energies. The resulting diagram, however, cannot be treated as if all states

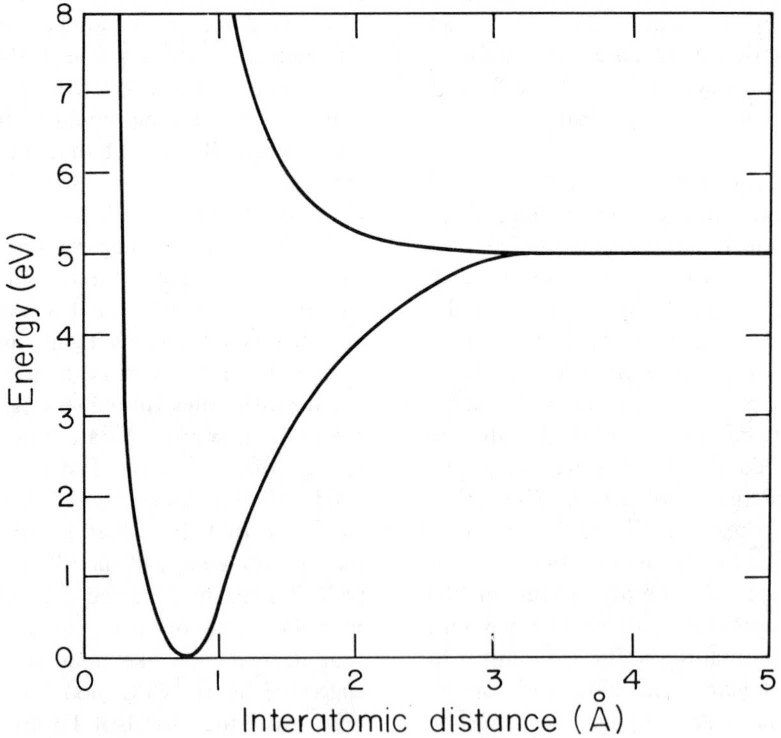

FIGURE 8. Energy as a function of interatomic separation for the lowest bonding ($^1\Sigma_g^+$) and antibonding ($^3\Sigma_u^+$) orbitals of a hydrogen molecule.

were ordinary one-electron states. In particular, the upper states are not accessible unless the lower ones are filled, and Fermi statistics are applied by treating the impurity states as a system in thermal and diffusive contact with the rest of the electronic system. Thus the Fermi energy of the material fixes the average ionization state of the impurities, but complicated statistics result from the interactions between localized electrons.

In an ordinary semiconductor, there are few states in the gap, and the Fermi energy, particularly at low temperatures is very sensitive to the concentrations and types of impurities or defects present. On the other hand, in the CFO model for covalent amorphous alloys, the large value of the density of states throughout the entire gap results in an effective pinning of the Fermi energy, independent of small changes in composition or defect concentration. If the density of states at the Fermi energy is $g(E_f) = g_c(E_f) + g_v(E_f)$, then introduction of ΔN additional electrons, via, for example, injection or creation of donor sites, results only in a small increase of the Fermi energy to $E_f + \Delta E_f$, where

$$\Delta E_f \approx \Delta N / g(E_f) \quad . \tag{26}$$

Since the conductivity in the CFO model is still given be Equation 24, a small change in the Fermi energy will not result in a conductivity jump of many orders of magnitude, as happens when a similar increase of electrons occurs in a crystalline semiconductor or even in an amorphous semiconductor with non-overlapping band tails. As we shall see, this consequence of the CFO model, although not unique to the model, predicts several of the unusual physical properties observed in amorphous semiconductors.

We have still neglected electronic correlations. As discussed in section II, correlations become important in localized states because the ionization potential of an atom or ion always exceeds its electron affinity. The effective difference between these quantities is the intraionic Coulomb repulsion that we called U in section II. Screening will reduce U considerably,[122] and the fact that the localization need not be of the order of an atomic radius further reduces its magnitude, but a reasonable estimate for U based on a multivalent donor analogy is of the order of 0.2 eV.[123] Since the crystalline solids corresponding to the covalent amorphous semiconductors under discussion appear to be adequately handled by one-electron theory, we can conclude that U is negligible for the delocalized states outside the mobility gap. Thus electronic correlations become important at the mobility edges. The resulting situation is extremely complex. One result of a finite U is a splitting of the localized states by a term of order U. Since states are only split when one of them is localized, the density of localized states depends sensitively on the occupation numbers. If we make the brutally oversimplified assumption that U is constant in the mobility gap and zero outside, an effective density of states such as that shown in Figure 9 could result. The major new feature of Figure 9 is that discontinuities in the density of states due to electronic correlations occur at the mobility edges.

The electron-phonon interaction has also been neglected. It is clear that, for example, the redistribution of electronic charge density inherent in the CFO model couples strongly at least to the optical phonons and thus induces ionic effects. In particular, if some of the normal modes of the material are localized, as is certainly possible in disordered systems, ionic distortions will occur in those regions in which the charge redistribution takes place. As discussed in section II, it is very likely that in this type of a situation small-polaron states are appropriate. If the binding energy of the small-polaron is sufficiently large, the effective value of U discussed previously is reduced.[123]

As we have noted, conduction in states above the conduction-band mobility edge and below the valence-band mobility edge should be band-like in nature, and does not need the assistance of phonons to proceed. Thus, for this range of energies, $\mu(E)$ remains finite, even at T=0. If we use the Kubo-Greenwood[116,117] formula, the mobility can be expressed[100,124]

$$\mu(E) = \frac{2\pi e\hbar^3 a\lambda^2}{3m^2 kT} g(E) \tag{27}$$

where λ is proportional to the momentum matrix element between nearest-neighbor localized states, and is of the order of unity or smaller.[124] Equation 27 can be used to place a lower limit on the mobility of delocalized states of the order of 5 cm^2/V-sec.[100,124,125] This is slightly higher than the estimate obtained in section II for the limit of a band-like mobility, based on a minimum mean free path of one interatomic spacing. Cohen[125] has suggested that conduction in the range below the limit of validity of Equation 27, where the

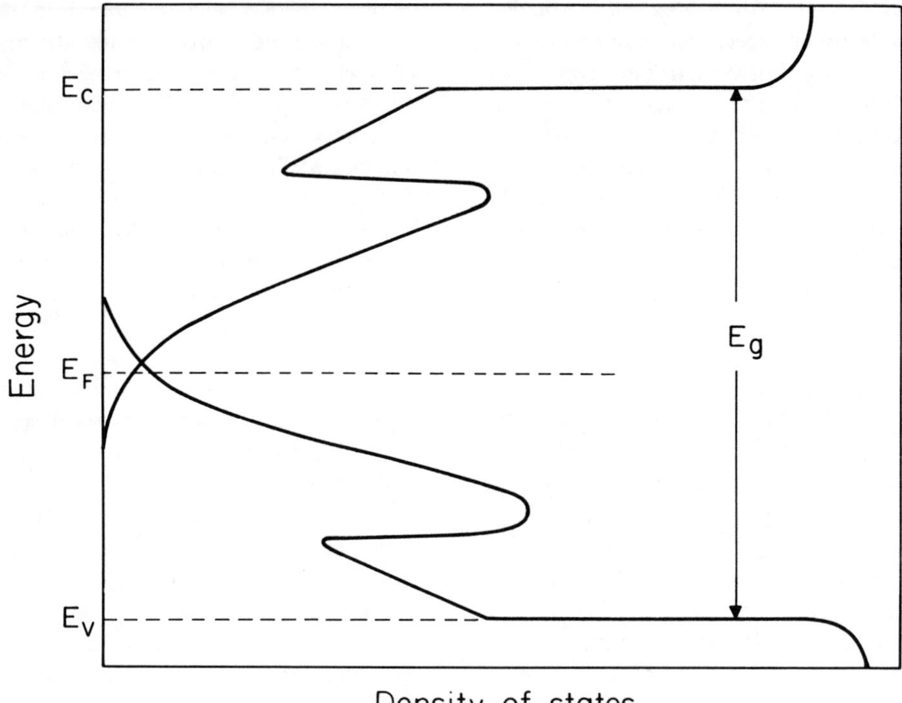

FIGURE 9. Effective one-electron optical density of states for a CFP-type amorphous semiconductor in which electronic correlations are important for localized states in the mobility gap.

mean free path is of the order of the interatomic separation, be more properly considered as a diffusive or Brownian-type motion, although the states remain delocalized.

On the other hand, conduction among states within the mobility gap takes place only with the assistance of phonons. The mobility then increases exponentially with temperature in accordance with Equation 15. As discussed previously, it is likely that small polarons form. However, because of the lack of long-range order, the small-polaron band regime will be absent, and conduction takes place only by means of hopping down to T=0. Mott[100,126] has pointed out that at very low temperatures, the predominant conduction mechanism will be phonon-assisted tunneling. This comes about because the localized electrons can find a state available farther away than on the nearest-neighbor atoms that has energy considerably closer to its own. The mobility for phonon-assisted tunneling must be of the form

$$\mu = Ae^{-\alpha R}\, e^{-W/kT}$$

where R is the tunneling distance, α is a constant depending on the barrier height, and W is the energy difference between initial and final states. For nearest-neighbor hopping, R=a, $W=E_a$, and we recapture the form of Equation 15. But if the electron hops to a farther neighbor, R=na, it can find a localized state with energy nearer its own by a factor of n^3; thus $W=E_a/n^3$. Then the mobility is the sum of terms of the form

$$\mu_n = Ae^{-(\alpha an + E_a/n^3 kT)}$$

The predominant term is the one for which the exponent is a maximum; this occurs for

$$n = (3E_a/\alpha akT)^{\frac{1}{4}}$$

At high temperatures, n may be of the order of unity or less, in which case the derivation is inapplicable and nearest-neighbor hopping predominates. But at sufficiently low temperatures, n is large compared to unity, and phonon-assisted tunneling to farther atoms becomes important. In this region, the mobility takes the form

$$\mu = Ae^{-B/T^{\frac{1}{4}}} \qquad (28)$$

where $B = 4\ (\alpha a/3)^{3/4}\ (E_a/k)^{1/4}$. Although crude,

Equation 28 can be used as a test of whether band-like or hopping conduction predominates in a particular material.

As mentioned in section I, we shall not be particularly concerned with phonons in this review. Nevertheless, the effects of disorder on the phonon spectra of amorphous solids should be analogous to those on the electronic spectra. It is clear that the normal modes of vibration cannot be described by the wave vector **k** and that the distinctions between transverse and longitudinal phonons and between acoustic and optical branches can no longer be made unambiguous. On the other hand, there should be no differences between the long-wavelength acoustic vibrations of corresponding amorphous and crystalline solids, since this is the region of propagating sound waves, in which it is valid to treat the material as a continuum. Entirely analogous to the electronic density of states, we can define a density of phonon modes that should be obtainable experimentally. We should expect that the effects of disorder should lead to a disappearance of the van Hove singularities and to an extensive "band tailing" in the density of phonon modes of an amorphous solid. It is possible that phonon modes deep in the tails are quite localized in space, particularly in the high-frequency region of the spectrum. However, as we cautioned in the Introduction, the density of phonon modes in corresponding crystalline and amorphous solids will resemble each other much less than the density of electronic states if greater than second-neighbor force constants are important.

It is of the utmost importance to calculate expressions for the experimentally observable transport and optical coefficients in each model for the electronic band structure of amorphous semiconductors, in order to test for the applicability of the model. For band-like conduction above a mobility edge, it can be shown[124,127] that the ordinary semiconductor expressions for the thermoelectric power, e.g. Equation 17, remain appropriate. The ac conductivity in this model should be essentially independent of frequency up through the microwave region. On the other hand, when hopping conduction predominates, the ac conductivity should increase roughly as $\omega^{0.8}$, as shown by Pollak and Geballe.[128] Mott[129,130] has derived an approximate expression for the ac hopping conductivity in the CFO model:

$$\sigma(\omega) = \frac{1}{3}\,\pi e^2 a^5\,[g(E_f)]^2\,\omega[\ln(\omega_o/\omega)]^4\,kT \qquad (29)$$

where ω_o is the optical-phonon frequency and a gives the spatial extent of the localized states in the vicinity of the Fermi energy. Mott[6] suggests that the thermoelectric power in this regime should be identical to the expression used for metals,

$$S = \frac{\pi^2 kT}{3e}\left[\frac{d\,\ln\sigma(E)}{dE}\right]_{E=E_f},$$

since the Fermi energy falls in a region where the density of states is finite. However, this relationship can be justified only within Boltzmann transport theory, which may not be applicable for conduction in localized states.

The optical absorption coefficient, $\alpha(\omega)$, can be evaluated from the ac conductivity at optical frequencies from the expression

$$\alpha(\omega) = \frac{4\pi}{nc}\,\sigma(\omega) \qquad (30)$$

where n is the index of refraction. For amorphous solids, **k** is not a good quantum number, and it is almost certain that the **k**- conservation rule breaks down completely, at least in all localized states and in delocalized states near the mobility edges. The absorption constant then can be written[103]

$$\alpha(\omega) = \frac{8\pi^4 e^2\hbar^2 a}{m^2 nc\omega}\int dE\; g(E)\; g(E+\hbar\omega)\;|M(E)|^2 \qquad (31)$$

where M(E) is an effective matrix element for the optical transition. As defined in Equation 31, $M \sim 1$ for transitions between delocalized states. Mott[130] suggested that M is of the same order of magnitude for transitions between a delocalized and a localized state. At first sight, this appears surprising, since two delocalized wave functions overlap in the vicinity of all N atoms, while a strongly localized wave function overlaps a delocalized wave function only in the vicinity of a single atom. However, this ignores the required normalization. If we write the delocalized wave functions as linear combinations of localized functions,

$$\phi_{n,m}(\underset{\sim}{r}) = \sum_l A_{nml}\,\psi(\underset{\sim}{r} - \underset{\sim}{R}_l) \qquad (32)$$

where n is a band index and R_l is an atomic position, then normalization should result in the amplitude of ϕ being smaller than that of ψ by a factor of $\sqrt{N}$. This would lead to a matrix element for a delocalized-localized transition smaller by a factor of $\sqrt{N}$ than that for a delocalized-delocalized transition. However, as pointed out previously, if we assume that the phases of the A_{nm} are random for delocalized states near the mobility edges, then the contributions to M $a\int$ dr $\phi^*_{n,m}$ (r) $\phi_{n,m}$ (r) from delocalized-delocalized transitions are reduced by an additional factor of $\sqrt{N}$, just sufficient to make the two types of matrix elements the same order of magnitude.[124,130] Thus, as long as the approximation of random phases is valid, no break will appear in $\alpha(\omega)$ at the mobility gap. However, this entire treatment neglects the effects of electronic correlations completely. But, as discussed previously, these must be important in localized states. In fact, Figure 9 suggests that the optical gap is sharp, and thus that correlations suppress the delocalize-localized transitions.

Although the matrix element, M(E), may not vary greatly when one of the electronic states involved is localized, the same will not be true when *both* are localized. As pointed out previously, states in the conduction-band tail are spatially well-separated from states in the valence-band tail. Thus, the wave functions of two localized states in different bands will overlap each other negligibly, and we can assume M≈O for localized-localized transitions. We still cannot evaluate $\alpha(\omega)$ from Equation 31 without a knowledge of g(E) for both valence and conduction bands, as well as the Fermi energy and the mobility edges. For crystalline semiconductors, $g(E)\alpha(E-E_o)^{1/2}$ near a band edge, but the presence of band tails in amorphous semiconductors destroys this simple relationship. Mott[100,103] has presented a simple argument that

$$g(E) \propto (E - E_o) \tag{33}$$

near the band edge of an amorphous solid with perfect short-range order. It is possible that imperfections in the short-range order invalidate this relationship over the entire range in which it has been derived, but it probably always remains a better approximation than that for crystalline solids. If E_{vo} and E_{co} represent the valence and conduction "band edges" respectively extrapolated from the linear regions given by Equation 33, then the absorption constant can be evaluated as

$$\alpha(\omega) \propto (\hbar\omega - E_{opt})^2/\hbar\omega \tag{34}$$

where E_{opt}, the optical gap, is either (E_c -E_{vo}) or (E_{co} -E_v), whichever is smaller.[130] Since (34) has been derived in such a crude manner, we should not expect it to be of general validity. In any event, the optical gap bears little relationship to any other important physical quantities. It can be easily obtained, however, from a plot of $(\alpha\hbar\omega)^{1/2}$ as a function of $\hbar\omega$ by extrapolating the linear region, if one is observed, to the $\hbar\omega$ axis.

The interpretation of other types of experiments, such as photoconductivity, tunneling, recombination radiation, etc., although implicit in the previous discussions of g(E) and μ(E), will be given together with the analysis of individual materials in sections IV-VI.

It should be borne in mind that although the models discussed in detail here provide evidence in favor of the concepts of band tails and mobility edges in disordered systems, without quantitative calculations, the extent of the band tails and the positions of the mobility edges for any real material is entirely unknown. It is possible that the band tails of amorphous solids are sufficiently small that the mobility edges are virtually indistinguishable from the ordinary band edges in periodic crystals. A finite density of states at the Fermi energy, g (E_f), can still result provided compensating donor and acceptor levels exist in the material with relatively large concentrations. Thus the Fermi energy can be pinned without overlapping band tails, and, as will become evident in the following two sections, it can be extremely difficult to experimentally distinguish such a model from the CFO model.

IV. ELEMENTAL AMORPHOUS SEMICONDUCTORS

A. Ge and Si

Ge and Si are the two most investigated crystalline semiconductors, and their similarities are in general more striking than their differences. At the present state of our knowledge, the same is true for their amorphous forms, and consequently it is appropriate to discuss them together. The major features of amorphous Ge and Si, as well as those of their crystalline forms, arise from the chemical bonding, which is completely covalent. Both materials have four outer electrons and thus each atom is in its lowest energy state when it is surrounded by a regular tetrahedron of like atoms, each covalently sharing one electron with the central atom. A regular array of such tetrahedra forms the diamond structure of crystalline Ge and Si. In order to obtain amorphous Ge and Si, a disordered structure must be produced.

Because of the fact noted in section I that the short-range order of liquid Ge and Si is metallic in nature, amorphous semiconducting Ge and Si cannot be obtained by rapid quenching from the melt. Such a procedure has thus far yielded only polycrystalline samples. Consequently, bulk amorphous material has not yet been produced. Amorphous films are commonly obtained by one of four main methods - vapor deposition,[131] sputtering in argon,[132] electrolitically from solutions of materials such as $GeCl_4$,[133] and by means of a glow discharge from silane or germane gas.[134] It should be borne in mind that the method and exact details of the film preparation are of the utmost importance in comparing the various experimental results.

1. Structure

The bulk of our knowledge of the structure of amorphous Si and Ge comes from x-ray and electron diffraction experiments. Radial density functions have been obtained,[10,135,136] and show that the atoms in the amorphous solids are surrounded by a tetrahedron of other atoms with an average separation essentially equal to that in the corresponding crystal. Several models for the structure have been proposed, based on these results.[137] A simple possibility is that the amorphous material is made up of microcrystallites of, say, 20 Å in extent.[138] Another model is based on the fact that two possible crystal structures, the diamond and the wurtzite lattices, have identical tetrahedrally coordinated nearest-neighbor environment. The second nearest-neighbor environment of these two structures differ, however, the diamond lattice having only chain-like or "staggered" six-membered rings and the wurtzite structure having one-quarter boat-like or "eclipsed" siz-membered rings (see Figure 10). Grigorovici and coworkers[139-142] pointed out that a five-membered ring, with only a slightly deformed tetrahedral angle is also possible, and that these can be built up to form icosahedral arrays in which twenty tetrahedron centers occupy the corners of a regular dodecahedron. They called such structures "amorphons", and proposed that a mixture of half diamond units and half amorphons could account for many of the properties of amorphous Ge. Independently, Coleman and Thomas[143] proposed a similar model for Si based on a 60% diamond-40% amorphon ratio. Polk[144] constructed a random-network model,[145] in which all atoms are surrounded by four covalent bonded atoms at the proper interatomic separation (i.e. no internal dangling bonds), and no bond-angle distortion greater than 20° exists. Polk found that such a structure could be extended to fill all space with 97% of the density of the periodic crystal. Finally, Breitling,[146] proposed

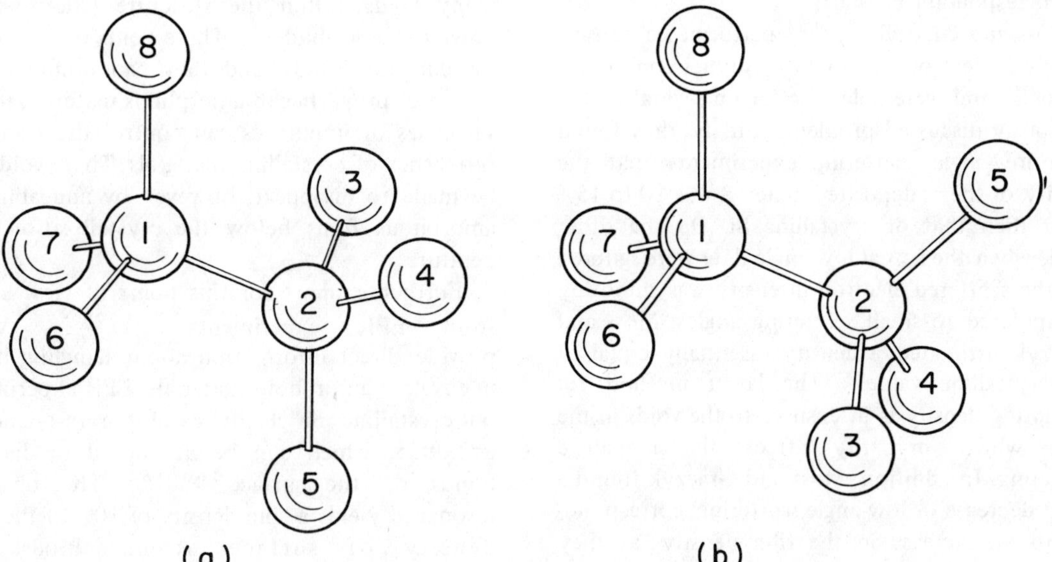

FIGURE 10. (a) Staggered (chain) configuration for tetrahedrally bonded structures. (b) Rotation of 180° necessary to obtain eclipsed (boat) configuration for tetrahedrally bonded structures.

two types of layered structures for amorphous Ge, based on distortions of the diamond lattice. He suggested that the first structure results when films are vapor deposited on a cold substrate, while the second structure, more closely connected with microcrystalline Ge, is obtained after annealing or storing the film for a long period of time.

Since all of these models are consistent with the early x-ray diffraction results, other experiments are necessary in order to discriminate between them. Such observations as film density, electron diffraction, and electron paramagnetic resonance (EPR) measurements have provided useful clues in this regard. The densities of amorphous Ge and Si are easily determined, and in both cases values of from 5 to 30% less than the corresponding crystalline densities have been observed. For amorphous Ge, Clark[147] reported a 28% decrease in density of his $\sim 20\mu$ vapor-deposited films. On the other hand, Light[148] found a density within 5% of the crystal, using an analysis based on x-ray absorption data. Donovan et al.[149,150] determined a 12 to 15% density decrease on relatively carefully prepared films. An entirely similar situation has been found for amorphous Si.[151,152] These results are not contradictory, since density is a macroscopic property, and includes voids, cracks, and pinholes, all commonly found in thin films.[153] Thus, the highest value obtained is the best guide to the ideal density of an amorphous material, and we can conclude that the densities of amorphous Ge and Si are quite close to those of the corresponding crystals.

Moss and Graczyk[152,154] performed extensive scanning electron diffraction studies on amorphous Si, and were able to clear up several of the previously discussed problems. Firstly, they found from low-angle scattering experiments that the density of the as-deposited material was 10 to 15% lower than that of crystalline Si. On the other hand, when the actual low-angle data were ignored and the scattered electron intensity was smoothly extrapolated to small scattering angles, Moss and Graczyk estimated a density essentially equal to the crystalline value. The latter method for estimating density is insensitive to the voids in the film, which primarily affect the low-angle scattering. In addition, Moss and Graczyk found a sharp decrease in low-angle scattering, corresponding to an increase in the film density, as they annealed their films below the crystallization temperature, strong evidence that the voids were vanishing with the heat treatment. Finally, from their observed radial density function, Moss and Graczyk determined that the first and second neighbor distances and coordination numbers for amorphous and crystalline Si are virtually identical, but that the third crystalline peak is essentially absent in the amorphous structure. Because of this feature of the data, no microcrystallite model of the large number they investigated could provide a fit to their intensity function. On the other hand, the random-network model of Polk[144] fits the data very well. Moss and Graczyk[154] also compared the *widths* of the peaks in the radial density function of a 100 Å as-deposited amorphous film with those of the same film after crystallization. Although the widths corresponding to the nearest-neighbor distances were approximately the same, the second-neighbor peak in the amorphous film was over twice as wide as that in the crystallized film. This increased linewidth corresponds to bond-angle distortions of as much as 20°, a result completely in accordance with the model of Polk.[144]

These structural results are consistent with the idea of an ideal amorphous state,[6,155,156] a state with near-perfect short-range order but no long-range order. Such a state has no internal dangling bonds, and all valence requirements in the interior are locally satisfied. This state is metastable, and represents a local energy minimum with respect to atomic rearrangements. As-deposited amorphous Ge and Si are far from the ideal state, and contain many voids within the structure (the so-called "Swiss cheese model"). These voids lead to interior dangling bonds, and they can dominate the electrical properties of amorphous materials just as vacancies or impurities can control the transport properties of crystalline materials. These voids can be made to disappear, however, by annealing the amorphous films below the crystallization temperature.

Further support for this point of view comes from EPR experiments,[136,157-159] which provide direct information about dangling bonds in covalent amorphous materials. EPR experiments on crystalline Si clearly exhibit a resonance at g=2.0055, which can be attributed to dangling bonds on the surface.[160-162] The observed resonance yields a spin density of 10% of the total density of surface atoms. Brodsky et al.[136,157-159] observed the resonance at g=2.0055 in amorphous Si films prepared by vapor deposition, rf sputtering and ion inplantation, and in each case found an unpaired-spin density of $2x10^{20}$ cm^{-3}, independent of film thickness. Assuming that the crystalline observation that only 10% of the actual dangling bonds are seen in an EPR experiment remains valid, Brodsky et al. concluded that the density of unpaired spins is $2x10^{21}$ cm^{-3} in amorphous Si, an extremely high value. The independence of this density on thickness implies that the dangling bonds are distributed throughout the entire film, rather than restricted to the surfaces. All this is consistent with the model of a distribution of voids throughout the as-deposited material. A density of $2x10^{21}$ cm^{-3} of unpaired spins correponds to an average separation of about 8 Å. Although this intuitively seems to be remarkably small, it is not an unreasonable value. Since the unpaired spins are concentrated at the inner surfaces of voids, rather than being randomly arranged, the concept of an average separation is misleading. If we assume that the voids are about 5 Å in size (about 20 missing atoms[137]), then an average distance between voids of 10 Å results in a 12% density deficiency, in the observed range. A spin density of $2x10^{21}$ cm^{-3} corresponds to just 2 dangling bonds per void, if anything, a value on the low side.

Brodsky et al.[136] also annealed their films below the crystallization temperature, and found that although the x-ray diffraction patterns of the films were indistinguishable, the EPR signal strength sharply decreased with annealing temperature. This implies that the voids anneal away, in agreement with the electron-diffraction results of Moss and Graczyk, Brodsky et al.[136] proposed an alternative "building block" model, based on a random arrangement of 10 to 15 Å microstructures, which could even be the amorphons of Grigorovici et al.[139-142] The major difference between the "building block" and the "Swiss cheese" models is that the former denies the existence of an ideal amorphous state. The annealing process can be thought of as a "shaking out" of the voids, as the blocks pack more densely. The concept of building blocks also necessitates the presence of sharp discontinuities between the microstructures, very similar to the grain boundaries of polycrystalline materials, and could manifest themselves in transport properties. Although it is difficult to distinguish between the "Swiss cheese" and "building block" models, particularly without restricting the nature of the blocks to a given structure, the former model explains the electron-diffraction results of Moss and Graczyk[152,154] in a much more straightforward manner than the latter. In the "building block" model, there appears to be no simple way to account for the 20° distortions in bond angles which are necessary to explain the observed width of the second-neighbor radial-density-function peak. If such large distortions are introduced into a particular microstructural unit such as an amorphon, then the result will be essentially indistinguishable from a random network.

Brodsky[163] recently reported that the EPR signal was not observed in amorphous Si prepared by means of an rf glow discharge from silane gas. This is evidence that dangling bonds are not necessarily present in all amorphous Si films. It is likely that Si prepared by this technique contains residual hydrogen, which congregates in the voids and saturates the potential dangling bonds. Further evidence for this conclusion comes from the transport data on such films. This focuses on an important problem. Until quantitative means are developed for analyzing the impurity content of amorphous Ge and Si films, we should be very cautious in attributing their physical properties to intrinsic behavior. For example, it is entirely possible that the annealing effects on vapor-deposited films results not from a coalescence of the voids, but rather from a diffusion of oxygen or other impurity into the voids, thus decreasing the dangling-bond density.

2. Transport Properties

Electrical conductivity in amorphous Ge and Si films has been studied often[134,136,139,147,164-185] and a consistent picture is finally beginning to emerge. Much of the older work was erratic in the sense that the actual values of conductivity at a given temperature differed by as much as a factor of 10^4, although plots of log σ vs. $1/T$ gave two straight-line regions, which were generally identified with "extrinsic" and "intrinsic" conduction. Typically, the low temperature activation energy for amorphous Ge was found to be 0.2 to 0.3 eV, and the high-temperature activation energy was 0.4 to 0.6 eV;[165-168] the corresponding activation energies are about 50% larger for amorphous Si.[164,169] Later work,[147,151,176,183] however, has shown that such plots are actually not straight lines and thus no true activation energy can be defined, at least below room temperature. Walley and Jonscher[176,178] were the first to show that the problem of measuring conductivity as a function of temperature was that the essential character of the films change with annealing. Thus σ (T) is reproducible only below the maximum temperature previously attained. As the structural studies discussed previously have shown, a large density of voids appear in the as-deposited material, and these anneal away with increasing temperature. This is shown clearly in Figure 11, taken from the work of Brodsky et al.[136] Annealing increases the resistivity (by a factor of 10^3 at 200K), primarily by increasing the apparent activation energy. This is easily understandable in terms of lowering the concentration of dangling bonds, in agreement with the EPR results. Conduction in the as-deposited film is extrinsic, and proportional to the density of dangling bonds, which can act either as donors or acceptors, depending on whether the

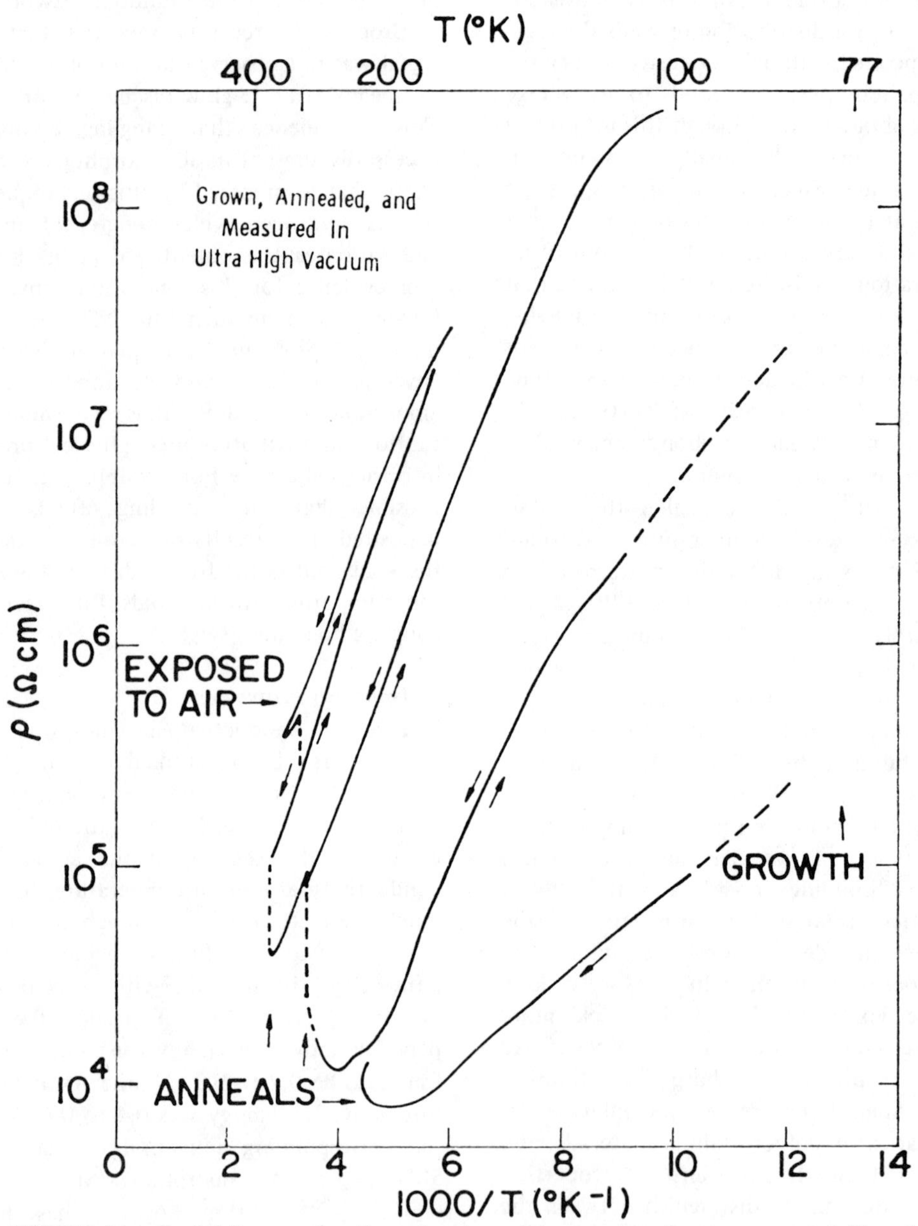

FIGURE 11. Resistivity as a function of reciprocal temperature for an evaporated film of amorphous Si, grown and measured in ultrahigh vacuums. The subsequent admitting of air is seen to produce only a small effect compared to the effects of annealing (Brodsky et al.[136]).

energy required to ionize the unpaired electron by placing it in the conduction band is smaller or larger than that required to saturate the bond by removing an electron from the valence band Figure 11 also shows clearly that the large resistivity increases on annealing are not due to oxygen contamination, since the effects were obtained in ultra-high vacuum. Exposure to the air resulted in only a small increase in resistivity.

It is tempting to conclude that the conductivity of the film obtained by annealing just below the crystallization temperature is that of ideal amorphous Si. The EPR signal has vanished at this point, implying that the dangling bonds have been eliminated. Brodsky et al.[136] find that such a film can be characterized by a conductivity of the form

$$\sigma \approx (0.002\ \Omega^{-1}\ \mathrm{cm}^{-1})\ e^{-(0.16\ \mathrm{eV})/kT}$$

in the 200 to 400K range. This is extremely unlikely to be the conductivity of intrinsic amorphous Si. The value of the pre-exponential is much too small and the activation energy is probably also too small to be intrinsic semiconduction. In fact, this conductivity, if extended to sufficiently low-temperature may indeed be better represented by the $T^{-1/4}$ behavior suggested by Equation 28. Stuke[182,186] has found a conductivity of the form

$$\sigma \approx (10^{4}\ \Omega^{-1}\ \mathrm{cm}^{-1})\ e^{-(0.80\ \mathrm{eV})/kT}$$

for amorphous Si films between 500K and 800K, and this type of behavior is more likely to represent intrinsic conduction, if the films have remained amorphous. For amorphous Ge in the 300 to 600K region, Stuke[182,186] observed

$$\sigma \approx (10^{4}\ \Omega^{-1}\ \mathrm{cm}^{-1})\ e^{-(0.56\ \mathrm{eV})/kT}$$

At lower temperatures, plots of log σ vs. 1/T are generally not straight lines. It is worthwhile to test whether the cause of this could be the predominance of thermally activated hopping conduction, in which case the mobility should have the behavior given in Equation 28. Remarkably, the data of Walley and Jonscher,[176] of Mogab and Block,[151] of Rusu et al.,[184] of Walley,[178] of Clark,[147] and Chopra and Bahl[183] all appear to be straight lines on log σ vs. $1/T^{-1/4}$ plots, over wide temperature ranges. Unfortunately, we cannot definitely conclude that hopping indeed predominates at low temperatures because of the crudeness of the theoretical derivation of Equation 28 and of the surprising fact that the experimental relationship is far from unique. For example, three amorphous Ge films of Walley and Jonscher[176] also obey the relation $\sigma = AT^{12}$ between 250K and 800K, and one film obeys $\sigma = BT^{7}$ in the 50 to 300K range. Similarly, Chopra and Bahl[183] find a proportionality of σ to T^{7} in the range 77 to 740K for an amorphous Ge film, and Clark's data[147] show that σ is proportional to $T^{8.3}$ from 25K to 200K on films of various thickness. Without any theoretical justification, these power-law dependences are merely curiosities, but they do caution us about the validity of the $T^{-1/4}$ fits. When data down to 4K become available, a better test of Equation 28 will be provided.

It has been well established[147,176,178] that conductivity is independent of film thickness. However, there can be a strong dependence of conductivity on the rate of deposition.[176,178] Walley[178] has shown that below a residual gas pressure of 10^{-5} torr and above a deposition rate of 50 Å/sec., the rate dependence of the conductivity disappears in either amorphous Ge or amorphous Si. At lower deposition rates or higher residual gas pressures, the resistivity increases sharply in both materials, indicative of oxygen contamination. Oxygen atoms can saturate the dangling bonds appearing on the surfaces of voids, locally forming compounds such as SiO, and thus decreasing the extrinsic conduction.

Amorphous Ge prepared by rf sputtering has been studied by Rusu et al.[184] who find that the resistivity of the purest films is essentially the same as that in evaporated films. Gas contamination provided films of substantially higher resistivity. Chittick et al.[134] found that amorphous Si prepared by an rf glow discharge from silane gas had resistivities as high as 10^{14} Ω-cm at room temperature, a factor of 10^{6} greater than measured in vapor-deposited films.[182] As previously mentioned, this is very likely due to the presence of residual trapped hydrogen in the voids of the film.

Perhaps a better test of whether or not hopping eonduction indeed predominates at low temperatures is from observations of the mobility. Attempts have been made to measure the Hall effect in amorphous Ge films, but the results are not in agreement. Piller and Khan[185] failed to detect any Hall voltage, and set an upper bound on the Hall mobility, as defined by Equation 16, of $\mu_H < 10^{-6}$ cm^2/V-sec. This result is difficult to accept, since it implies a free carrier concentration greater than 6x10^{21} cm^{-3}. On the other hand, Clark[147] observed an *n-type* Hall effect, which corresponded to a free carrier concentration of n~10^{18} cm^{-3} and a Hall mobility of 10^{-2} cm^2/V-sec at room temperature. The Hall mobility appeared to increase with increasing temperature in the 230 to 300K range, but the errors at low temperature were sufficiently large that no definite conclusion could be drawn. Nwachuku and Kuhn[187] observed a Hall mobility of 0.03 cm^2/V-sec and a carrier concentration of about 10^{18} cm^{-3}, in substantial agreement with Clark's data, except that the Hall effect was p-type. There is clearly a need for more accurate Hall measurements at this time.

Thermoelectric power experiments are simpler to perform, but much more difficult to interpret. Grigorovici et al.[175,188,189] measured the Seebeck coefficient in amorphous Ge and Si films, and found that at higher temperatures, but below the crystallization temperature, the thermoelectric power, s, approached the values in the corresponding intrinsic crystals. At intermediate, 250 to 500K, the behavior of s indicated p-type conduction. Assuming that the semiconductor occurred in only a single parabolic band, the mobility of carriers in both amorphous Si and Ge could be estimated as of the order of 10^{-2} cm^{-2}/V-sec. The corresponding free carrier concentrations were 10^{18} cm^{-3} in Ge and 10^{16} cm^{-3} in Si. Except for a problem in signs, these results are consistent with those obtained from Hall measurements. Below 250K, the sign of the thermoelectric power of amorphous Ge reversed, and its magnitude became small. This sign change could indicate the onset of impurity conduction, as is common in crystalline materials. The thermoelectric power of crystalline Si became n-type *above* 500K. Chopra and Bahl[183] also found a p-type thermoelectric power in amorphous Ge films in the 300 to 400K region, with a reversal in sign below room temperature. To compound the sign problems, however, Stuke and co-workers[181,182,186] observed an *n-type* thermoelectric power between 200K and 650K, with a maximum in magnitude near 400K. Below 200K, a sign reversal to p-type behavior occurred. Stuke[181] interpreted this as extrinsic hole conduction at low temperatures, transforming above 200K to intrinsic conduction in which the electron mobility exceeds the hole mobility. The qualitative behavior of S was similar to that of p-type crystalline Ge, except that the transition to the intrinsic region was much more gradual in the amorphous films. A possible explanation for the lack of sharpness is that the shift of the Fermi energy of amorphous Ge with temperature is very slow, due to localized states in the band tails. Stuke[181] suggested that the lack of a definite activation energy in the electrical conductivity is further evidence for a temperature variation of the Fermi energy.

A more direct method for determining the mobility of carriers is by measurement of the transit time between the electrodes of photoexcited carriers.[190] Such drift mobility experiments were performed by LeComber and Spear[191] on amorphous Si films prepared by rf glow discharge decomposition of silane. As previously indicated, such films have anomalously high resistivity, probably due to the trapping of residual hydrogen in the voids of the material. The conductivity as a function of temperature showed three distinct regions. Below 200K, the conductivity was of the order of 10^{-15} Ω^{-1} cm^{-1} and varied only slowly with temperature. The activation energy was ~0.1 eV, and a $T^{-1/4}$ dependence could not be ruled out in the temperature range investigated. Between 200K and 240K, the conductivity increased exponentially with an activation energy of 0.51 eV. Above 240K, the activation energy appeared to increase to 0.62 eV. The observed drift mobility also increased with increasing temperature. Below 170K, the mobility was about 0.002 cm^2/V-sec., and it varied relatively slowly with temperature. Between 170K and 240K, the drift mobility was characterized by an activation energy of 0.09 eV. Above 240K, the activation energy of the mobility increased to 0.19 eV. The room-temperature drift mobility was about 0.02 cm^2/V sec.

A basic difficulty in interpreting drift-mobility experiments in the band regime is that if trapping and thermal release occur often during transit, the measured transit time includes the time the carrier is trapped in a localized state, and the observed drift mobility is then significantly lower than the conductivity mobility.[192] Since this is almost undoubtedly the case for amorphous semiconductors, which have a continuous distribution of localized states in both the valence and conduction band tails, evaluation of the actual mobility of carriers in delocalized states becomes difficult. Nevertheless, an immediate result of the observed room-temperature drift mobility is that the free-carrier mobilities estimated from the thermoelectric power and Hall data are undoubtedly much too small.[193] LeComber and Spear assumed that the activation in the 240 to 300K region was due to thermal excitation of trapped carriers. From this, they could estimate a band tail of the order of 0.2 eV. The further assumption of a linear increase in the density-of-states of the band tails led to a free-carrier mobility of 10 cm^2/V-sec.

The ac conductivity is independent of frequency up to $4x10^4$ Hz at room temperature;[176] at higher frequencies, an increase approximately proportional to $\omega^{0.8}$ is observed.[183] This is the frequency dependence expected for hopping conduction and is consistent with observations on several other amorphous semiconductors. However, as opposed to these other materials, the ac conductivity of amorphous Ge increases strongly with temperature.

Electron tunneling experiments have been performed on amorphous Ge[187,194] and amorphous Si.[195] Nwachuku and Kuhn[187] reported sharp increases in the tunneling current in amorphous Ge at bias values differing by about 1V, but these results have not been reproduced.[194,196] On the other hand, Osmun and Fritzsche[194] and Sauvage et al.[195] obtained monotonic increases in conductance with increasing magnitude of bias, the conductance being essentially independent of polarity. This latter symmetry implies that the Fermi energy is near the center of the mobility gap, a result thus far unexplained. Although this is consistent with a band structure in which the valence and conduction bands tail symmetrically, there is no obvious reason why this should be the case in both Ge and Si. Osmum and Fritzsche pointed out the alternative possibility that the tunneling takes place into a surface layer with vastly different band structure than the bulk of the material.[197]

Although the tunneling impedance of the oxide is much larger than the impedance of the amorphous semiconductor in the 77 to 300K range investigated, the conductance at low biases is extremely temperature dependent, exhibiting approximately a $e^{-A/T^{1/3}}$ behavior at zero bias.[195] This is, at first sight, surprising, since the tunneling conductance through the high-impedance oxide is essentially independent of temperature. However, it is clear that at low biases, the tunneling is into localized states in the semiconductor. Thus these states do not have amplitudes at the oxide-semiconductor interface, and the tunneling distance is larger than the thickness of the oxide. The conductance is thus proportional to

$$G \propto e^{-(\alpha d_o + \alpha' d_s)}$$

where a and a' characterize the oxide and semiconductor barriers, respectively, d_o is the thickness of the oxide, and d_s is the smallest distance between a localized state with an energy within kT of that of the tunneling electron and the oxide-semiconductor junction. But since there are $g(E_f)$ kT localized states of the correct energy per unit volume in the semiconductor,

$$G \propto e^{-\alpha d_o} e^{-A/T^{\frac{1}{4}}}$$

Thus the temperature of the conductance reflects the fact that the electrons must tunnel farther into the semiconductor at low temperatures than at high temperatures to find a state within kT of the Fermi energy. This is an important result since it is direct evidence of a finite density of localized states near the Fermi energy of the amorphous semiconductor. Furthermore, since the temperature dependence of the tunneling conductance parallels the conductivity of the amorphous material itself, this is strong evidence for the predominance of phonon-assisted tunneling conduction in the as-deposited films, at least in the 77 to 300K range. The temperature dependence of the ac tunneling conductance and the ac conductivity of amorphous Si are also parallel,[198] further support for this conclusion.

For biases above 0.8V in amorphous Si, the tunneling conductance becomes temperature independent,[195,198] evidence that the mobility gap in the material is approximately 1.6 eV. This is consistent with the value estimated from the intrinsic conductivity measurements of Stuke[182,186] and some of the optical absorption results. The valence and conduction band tails estimated from tunneling experiments[195] appear to be exponential and quite symmetric, the Fermi energy falling in the center of the mobility gap. No structure could be detected in the gap, in agreement with the CFO model, but not the model of Davis and Mott.[130]

Another technique for investigating the density of localized states in the gap of amorphous Ge was used by Kastner and Fritzsche,[199] who studied the conductance changes of 100 to 2500 Å evaporated films upon absorption of water, ozone, ammonia, and ethanol on the surface of the material. Chemisorption on solid surfaces is always accompanied by some positive or negative charge transfer to the solid. Some of this charge is trapped in surface states, some more is trapped in localized states in the bulk, and if the density of localized states is too small to accommodate the entire space charge, the rest enter delocalized states beyond the mobility edge and contribute to electrical conductivity parallel to the surface. Kastner and Fritzsche calculated the fractional change in conductance expected from a given amount of charge transfer, assuming a density of states at the Fermi energy, $g(E_f)$, as suggested by the CFO model described in section III. In general, an increase in conductance should occur for one sign of charge transfer, a decrease for the other. However, if both electrons and holes contribute to the current flow, a minimum in the curve of fractional conductance as a function of charge transfer exists. Experimentally, it has been found that water, ethanol, and ammonia transfer electrons to Ge when chemisorbed, but that ozone takes electrons from Ge.[200] Kastner and Fritzsche[199] observed an increase in conduction upon passing any of these four vapors over amorphous Ge films of various thickness. In conventional terms, this can be explained only by the conclusions that both electrons and holes participate in conduction and that the Fermi surface of the films is initially always near the minimum in the conductance curves. There appears to be no obvious reason why the latter should be true. Kastner and Fritzsche further observed conductance changes of films of several different thicknesses when varying rates of water vapor were passed over the surface, and found that the conductance failed to saturate, neither with film thickness nor rate of flow. They were forced to conclude that water molecules do not chemisorp only on the surface, but also diffuse into the films and are absorbed internally, most likely on the inner surfaces of the voids discussed previously.

Piezoresistance experiments have been carried out on both amorphous Ge and amorphous Si.[201-204] Devenyi et al.[201-203] found evidence for three distinct temperature regions in both materials. At low temperatures (below 230K in amorphous Ge, below 400K in amorphous Si) the elastoresistance coefficient, $M \equiv \Delta\rho/\rho\,\epsilon$, where ϵ is the applied uniaxial strain, was essentially independent of temperature. The magnitude of M was small, and the sign was negative for amorphous Ge and the longitudinal coefficient of amorphous Si; the transverse coefficient of amorphous Si was found to be positive. There is no question, from the point of view of either the magnitudes or the temperature dependence, that the piezoresistance represents only a minor effect.[205] This is opposed to the major effects observed in crystalline Ge and Si.[206,207] The low-temperature regions were interpreted by Devenyi et al.[201-203] to be regions in which hopping conduction predominates. However, if conduction were primarily via a hopping mechanism, we should expect large, positive values of M, since the mobility must be very sensitive to interatomic separation. The hopping mobility can be expressed as[100]

$$\mu = (e v_{ph} a^2 \phi/kT)\, e^{-E_a/kT}\, e^{-2\alpha a} \qquad (35)$$

where v_{ph} is an average optical-phonon frequency, ϕ is a constant, E_a is the hopping activation energy, α gives the rate of fall-off of the localized wave functions, and a is the average interatomic separation. It is clear that the mobility should increase very rapidly with decreasing a unless the wave functions are quite spread out (small α). Thus the piezoresistance results are not consistent with a hopping model in the low-temperature region. Recently, Fuhs and Stuke[204] also obtained a small, negative temperature-independent elastoresistance in this temperature range in amorphous Ge prepared by electron-beam evaporation.

At high temperatures[203] (above 400K for amorphous Ge, above 550K for amorphous Si), the elastoresistance coefficient is negative and appears to be proportional to T^{-1}, characteristic of a major effect.[208] The magnitude of the elastoresistance increases with temperature for amorphous Si but decreases for amorphous Ge. The most likely explanation of these results is a change of the conduction-band mobility edge with applied stress. Assuming that this is responsible for the entire effect, it can be concluded that the mobility edge in amorphous Ge increases with stress relative to the Fermi energy by 0.8×10^{-6} eV/bar, while that of amorphous Si decreases by 1.0×10^{-6} eV/bar. The corresponding single crystal values are an increase of 2.5×10^{-6} eV/bar for Ge and a decrease of 0.8×10^{-6} eV/bar for Si.[209] Thus the signs of the edge shifts are the same in corresponding amorphous and crystalline materials, although the magnitudes differ. There is no particular reason why we should expect a strong correlation between these values, since they represent different effects. For one thing, the anisotropy of the crystalline bands is absent in the amorphous materials. Furthermore, in crystals, the activation energy in the intrinsic region is half the energy gap; the piezoresistance results give the stress dependence of this gap directly. On the other hand, as was pointed out in the discussion following Equation 24, the activation energy in an amorphous semiconductor is the difference between the conduction-band mobility edge and the Fermi energy, which is not necessarily in the center of the mobility gap; under an applied stress there will be not only a change in the relative separation of the valence and conduction bands, but also a significant shift of the individual mobility edges relative to the band edges. Furthermore, since the extent of the band tails themselves must be stress dependent, the mobility edges should also move relative to the Fermi energy. Since the bands increase in width with applied stress due to the increased overlap between nearest-neighbor wave functions, the band tails should become more intensive, thus bringing about a greater decrease in the conduction-band mobility edge than in the corresponding crystalline band edge. This effect should tend to decrease the stress coefficient relative to the crystal in both amorphous Ge and amorphous Si, and thus this could be a manifestation of an excess shift in mobility edge with stress. On the other hand, Fuhs and Stuke[204] measured a *positive* stress coefficient of the activation energy of amorphous Si of 0.1×10^{-6} eV/bar. Their corresponding result for amorphous Ge, a stress coefficient of 0.4×10^{-6} eV/bar, is in qualitative agreement with that of Devenyi et al.,[202] but is smaller by a factor of 2. Fuhs and Stuke also point out the large differences between corresponding coefficients in single crystalline and polycrystalline samples. These discrepancies underline the dangers in interpreting the piezoresistance results.

In both amorphous Ge and amorphous Si, intermediate temperature regions exist, in which the elastoresistance coefficients decrease with increasing temperature.[201-203] Since no major effects are evident in the resistivity results in these regions, there is no particular reason why the piezoresistance behavior need be interpreted as anything but a smooth transition from one conduction process to another. Some piezoresistance data are shown in Figure 12.

Magnetoresistance experiments on amorphous Ge and Si have recently been performed by Mell and Stuke.[210] As opposed to the results on corresponding polycrystalline films, the magnetoresistance of the amorphous films was negative and independent of the angle between the applied magnetic field and the electrical current. Furthermore, the magnitude of the magnetoresistance increased with field as $H^{0.5-1.0}$, rather than as H^2. A maximum in the magnitude of the magnetoresistance occurred at about 200K, and at low temperatures, for sufficiently low magnetic fields, a sign reversal to a positive magnetoresistance occurred. The magnitude of the magnetoresistance decreased with increasing temperature. Doping amorphous Ge with As or Ga did not affect the magnetoresistance, but 1% Sb caused the temperature of the sign reversal to increase. The magnetoresistance data for pure amorphous Ge are shown in Figure 13. The results for amorphous Si are similar to those for amorphous Ge.

Mell and Stuke interpreted the negative magnetoresistance as indicative of a hopping mechanism. An applied magnetic field should align the spins associated with the dangling bonds in amorphous Ge and Si, and thus increase the hopping probability. Indeed, the magnetoresistance observed in the impurity conduction of crystalline Ge and Si is generally negative.[211,212] Unfortunately, it remains negative above the critical concentration for the Mott transition to metallic conduction and even well into the region in which the Fermi energy enters the conduction band.[213] Attempts have been made to explain this effect by assuming the existence of localized moments,[214,215] but this model is not in agreement with experiment[216] and the negative magnetoresistance in degenerate crystalline semiconductors is still regarded as basically unexplained. Nevertheless, its existence cautions us against the unambiguous association of such a magnetoresistance behavior with hopping.

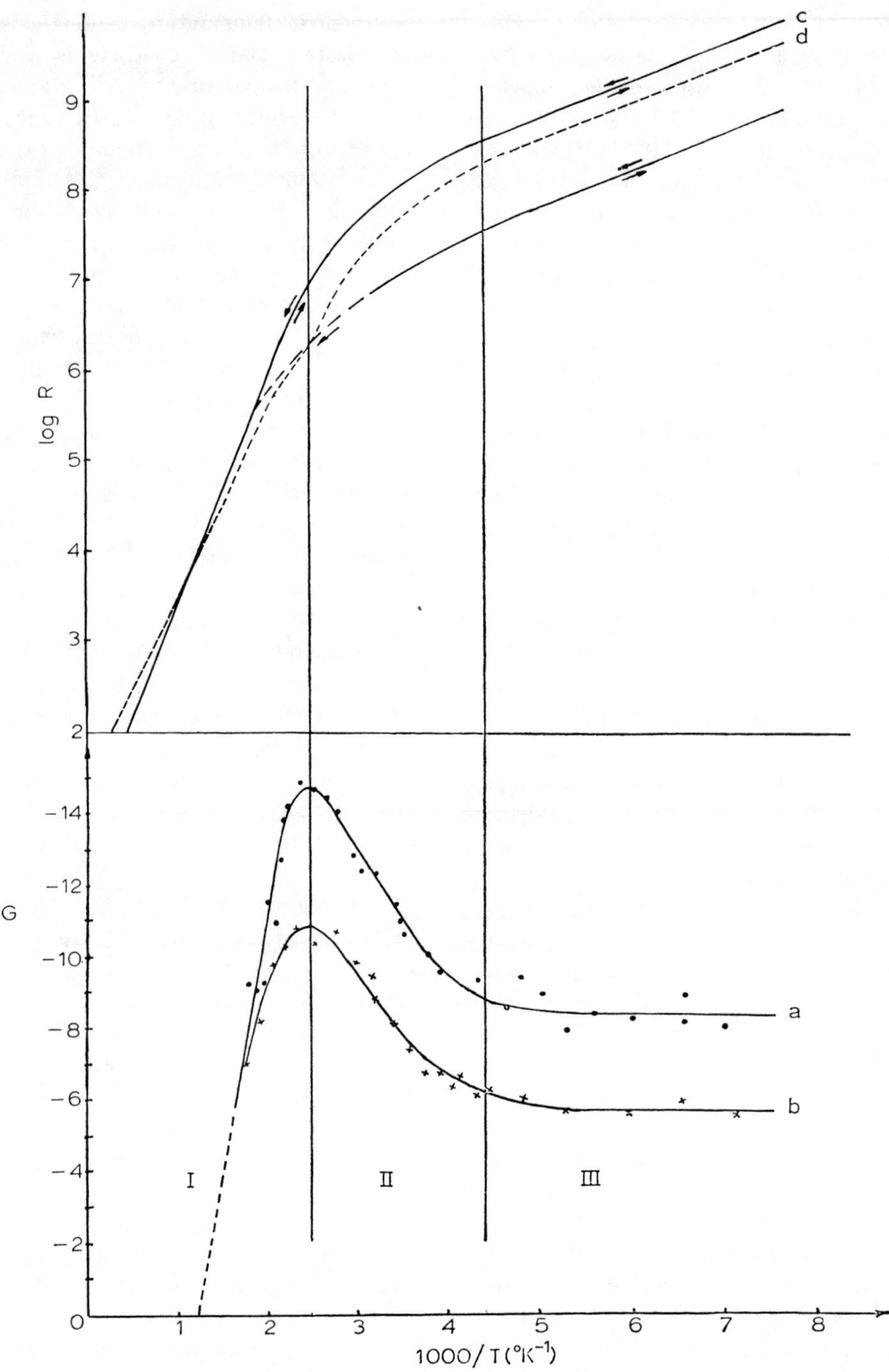

FIGURE 12. Results of piezoresistance measurements on amorphous Ge films. (a) Longitudinal elastoresistive coefficient. (b) Transverse elastoresistive coefficient. (c) Resistivity as a function of temperature at zero strain. (d) Resistivity as a function of temperature under a strain of 0.05, calculated from curve (a) (Devenyi et al.[202]).

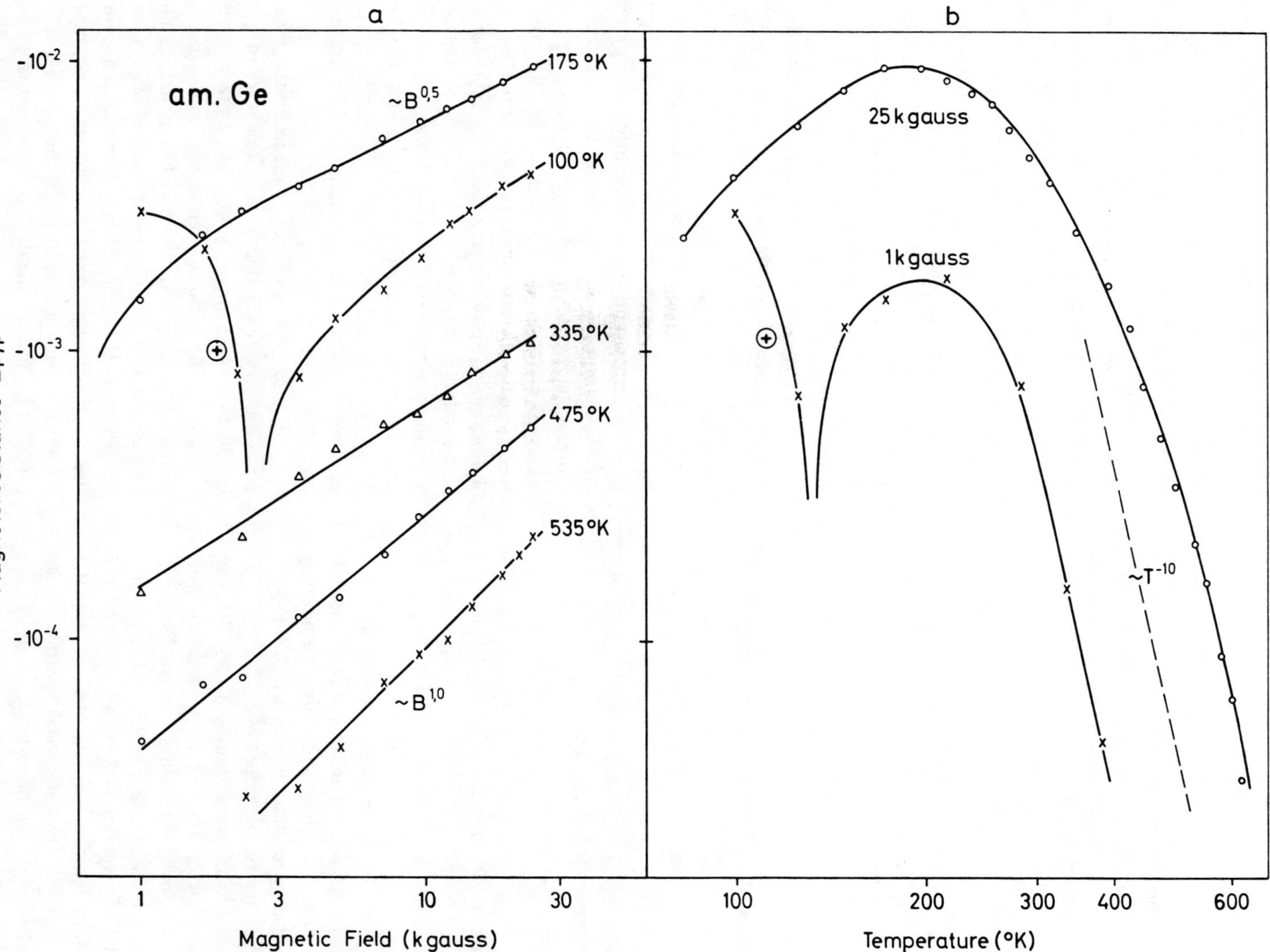

FIGURE 13. (a) Magnetic-field dependence of the magnetoresistance of pure amorphous Ge. (b) Temperature dependence of the magnetoresistance of amorphous Ge (Mell and Stuke[210]).

On the other hand, assuming that parallel hopping and band-like conduction mechanisms exist in amorphous Ge and Si provides a simple explanation for the magnetoresistance data. If we write

$$\sigma = \sigma_b + \sigma_h \tag{36}$$

where σ_b is the band contribution and σ_h is the hopping contribution, and further assume that $\sigma_b >> \sigma_h$ in the temperature range investigated, then the magnetoresistance is

$$\frac{\Delta\rho}{\rho} = -\frac{\Delta\sigma}{\sigma} = -\frac{\Delta\sigma_b + \Delta\sigma_h}{\sigma_b} \tag{37}$$

If the major effect of the applied magnetic field is to increase the hopping mobility, $\Delta a_h >> \Delta a_b$ and Equation 37 becomes

$$\frac{\Delta\rho}{\rho} = -\frac{\Delta\sigma_h}{\sigma_b} \tag{38}$$

In the 400 to 600K temperature range, Stuke[182,186] observed

$$\sigma_b = (10^4\ \Omega^{-1}\ \text{cm}^{-1})\ e^{-(0.56\ \text{eV})/kT} \quad .$$

In the same region, Mell and Stuke[210] measured the magnetoresistance at 25 kg as

$$\frac{\Delta\rho}{\rho} = -(5 \times 10^{-8})\ e^{(0.4\ \text{eV})/kT} \quad .$$

Putting these results into Equation 38 yields

$$\Delta\sigma_h = (5 \times 10^{-4}\ \Omega^{-1}\ \text{cm}^{-1})\ e^{-(0.16\ \text{eV})/kT} \quad .$$

Since there is no particular reason why the magnitude of $\Delta\sigma_h/\sigma_h$ should be a strong function of temperature, we can conclude that the hopping activation energy for amorphous Ge is of the order of 0.16e V, a reasonable value.

To summarize, it is apparent that very little can be concluded with any degree of certainty with regard to the conduction mechanisms in amorphous Ge and Si, despite the existence of a large body of experimental results. One of the major reasons for this is the inconsistency of the available data. The recent annealing studies have shown that it is of vital importance to subject the material to heat treatment to try to approach as close as possible to the ideal amorphous state to obtain reproducible results. In addition, as Bosnell and Voisy[217] have pointed out, the use of non-refractory electrodes lowers the crystallization temperature to values of the order of 100 to 300C in amorphous Ge and Si, and virtually guarantees crystallization near the contacts when electrical conductivity is measured. The crystallization creates barriers which lower the apparent conductivity upon annealing the films. There are still other dangers, such as gas inclusion, which is particularly evident in amorphous Ge and Si prepared by glow discharge or sputtering techniques. Consequently, it is not even clear that the intrinsic conductivity of these materials has yet been measured. It can be concluded from the structural studies and the model of Polk[144] that an ideal amorphous state, with no dangling bonds, is possible, although we cannot say how close we have come experimentally to achieving this state. The EPR studies, however, indicate that careful annealing below the crystallization temperature is a first step toward attaining such a state.

Provided there is a low impurity concentration and no important crystallization occurs near the electrodes, conductivity studies on carefully annealed films should provide the best indication of the intrinsic conduction processes in amorphous Ge and Si. Brodsky et al.[136] observed a low pre-exponential and a small activation energy on annealed films of amorphous Si, and this could represent intrinsic hopping conduction among localized states in the band tails. On the other hand, the high-temperature results of Stuke[182,186] possibly represent intrinsic band-like conduction, although it is difficult to explain why the pre-exponential factors for amorphous Si and Ge in these experiments are identical with those of crystalline Si and Ge despite the apparently much smaller mobilities of the former materials.

Hall and thermopower measurements have provided very few insights into the nature of the conduction mechanisms as yet, and more work is essential. The magnetoresistance and piezo-resistance results favor the conclusion that band-like conduction dominates at least above room-temperature. However, there is ample evidence for the existence of hopping conduction. The most convincing experiments in this regard are the drift mobility measurements of LeComber and Spear.[191] They found that the drift mobility of glow-discharge–produced amorphous Si was thermally activated, the activation energy being 0.09 eV below 240K and 0.19 eV above 240K. In the former range, the conductivity exhibited an activation energy of 0.51 eV, and this increased to 0.62 eV above 240K. It has been recently pointed out by Mort[218] that the fact that the discontinuities in the activation energies occur at the same temperature strongly indicate that the low-temperature conduction is primarily a hopping process, since the activation energy of a trap-controlled mobility does not appear in the conductivity.[219] Thus, it can be concluded that the hopping activation energy for glow-discharge-produced amorphous Si is 0.09 eV and that hopping occurs in localized states 0.42 eV above the Fermi energy. Above 240K, the predominant conduction mechanism is then band-like, the bottom of the band being 0.62 eV above the Fermi energy or 0.20 eV above the localized states in which the hopping takes place. The drift-mobility activation energy of 0.19 eV in this region identifies the trapping levels as just these localized states. It is not unreasonable to assume that there is a band tail in amorphous Si extending approximately 0.2 eV below the mobility edge. However, another possibility in the glow-discharge-produced films is that included hydrogen is responsible for a well-defined trapping level 0.2 eV below the band edge, and that the intrinsic band tail is much smaller in extent.

3. *Optical Properties*

In crystalline semiconductors, most of the insight into the intrinsic band structure has come from optical studies, primarily because of the relative insensitivity of the optical properties to impurities and defects, as compared to the transport properties. Consequently, it might be hoped that optical experiments on amorphous Ge and Si are more consistent and more easily interpretable than electrical conductivity results. To an extent, this is the case, but, for both materials, some inconsistencies are prevalent even in optical measurements.

There are now many optical absorption and reflectivity studies of both amorphous Ge and Si films.[136,147,149,150,183,185,220–232] Early work focused on comparisons of the reflectivity and of the imaginary part of the dielectric constant, ϵ_2, for corresponding amorphous and crystalline materials.[220,222] These results for Ge are shown in Figure 14. The two crystalline peaks near 4.5 eV and 6.0 eV are completely absent in the amorphous films. However, the third major peak of the crystal, near 2.2 eV, remains in amorphous Ge, but is very much broadened. Since the 2.2 eV peak has been associated with transitions in the (111) crystalline direction and the 4.5 eV peak with transitions in the (100) direction,[233] Stuke[181] suggested that the near-perfect short-range order in the amorphous material preserves the band structure along the (111) or nearest-neighbor direction. More recent work[150,228] has substantially confirmed the behavior originally found in ϵ_2, a single broad peak centered around 2.8 eV being the only structure observable in amorphous Ge. Analogous measurements on amorphous Si[229] have indicated that ϵ^2 is also quite smooth and has a single peak near 3.1 eV. As was the case for Ge, this 3.1 eV peak in Si[234] represents the effects of transitions in the crystalline (111) direction.[235] Tauc et al.[220,222] originally observed some structure below 0.6 eV in ϵ_2 and interpreted it as evidence that the spin-orbit-split valence bands of crystalline Ge are preserved in the amorphous material, but this structure has proved to be irreproducible.[150,236] Tauc et al.[236] have recently reported an absorption band at 0.23 eV in electrolytically produced amorphous Ge. It was much weaker in sputtered films and was absent in evaporated films. Thus, it does not appear to be intrinsic to amorphous Ge.

Less agreement exists over the nature and position of the absorption edges of amorphous Ge and Si. The early data[147,221,222,224,226] indicated an exponential tailing of the absorption coefficient at low energies, and was interpreted in terms of valence and conduction band tails arising from the positional disorder. No distinct values could be obtained for the energy gaps, as opposed to the situation in the corresponding crystalline materials, where a sharp fall-off in absorption is observed at particular energies.[237,238] This entire trend was reversed by measurements of Donovan et al.,[149,150] who obtained an absorption edge near 0.6 eV in carefully prepared films of amorphous Ge, the edge being essentially just as sharp as the direct edge in crystalline Ge. A similarly sharp edge was subsequently observed by Chopra and Bahl[183] near 0.5 eV. In both cases, the energy gap appeared to be smaller in amorphous than in crystalline Ge. On the other hand, Tauc et al.[236] and Beaglehole and Zavetova[229] have failed to reproduce the sharp edges of Donovan et al.[149,150] in amorphous Ge.

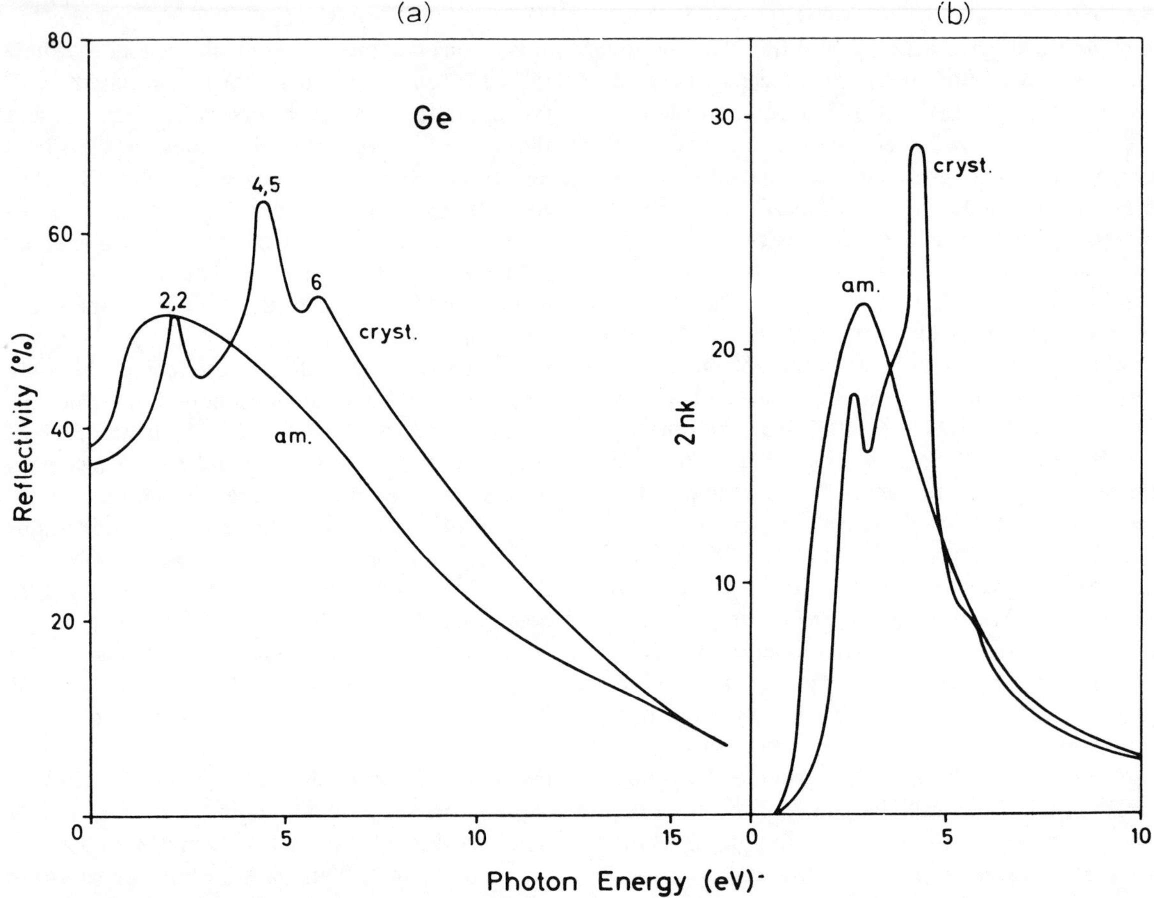

FIGURE 14. (a) Reflectivity of amorphous and crystalline Ge. (b) Imaginary part of the dieletric constant E_2 = 2 nk of amorphous and crystalline Ge (Stuke[181]).

The results of several sets of measurements are shown in Figure 15. There is no question that optical absorption is a very sensitive function of the quality of the films, and that this is responsible for the dispersion in results. Some light has recently been shed on this point by the experiments of Theye,[230] who found an exponential tailing in as-deposited films, but obtained a sharp fall-off in absorption after annealing the films. The apparent edge shifted to higher energies with continued annealing, and an anneal just below the crystallization temperature exhibited an energy gap of 1.0 eV, higher than that of crystalline Ge. After crystallization, the edge shifted to lower energy, in conflict with the results of Donovan et al.,[149,150] but in agreement with recent experiments by Paul et al.[232] Resistivity measurements by Theye[230] indicated that the highest quality films, judged by the smallest density of dangling bonds, were those with sharp absorption edges near 1.0 eV. These results are shown in Figure 16.

Connell and Paul[231] measured the pressure coefficient of the absorption edge of both evaporated and electrolytically deposited films of amorphous Ge and found identical shifts of about 3×10^{-6} eV/bar. This value cannot be correlated with the pressure coefficients of any of the critical points of crystalline Ge. The indirect gap in crystalline Ge shifts by 5×10^{-6} eV/bar[238] and the direct edge by 14×10^{-6} eV/bar.[240]

Once again, the situation with regard to amorphous Si appears to be entirely analogous to that of amorphous Ge, only with fewer data available. Grigorovici and Vancu[226] and Beaglehole and Zavetova[229] found exponential tails in their as-deposited films, the apparent edge shifting to lower energies than in crystalline Si. However, Brodsky et al.[136] observed a relatively sharp edge after annealing their films, the edge continually

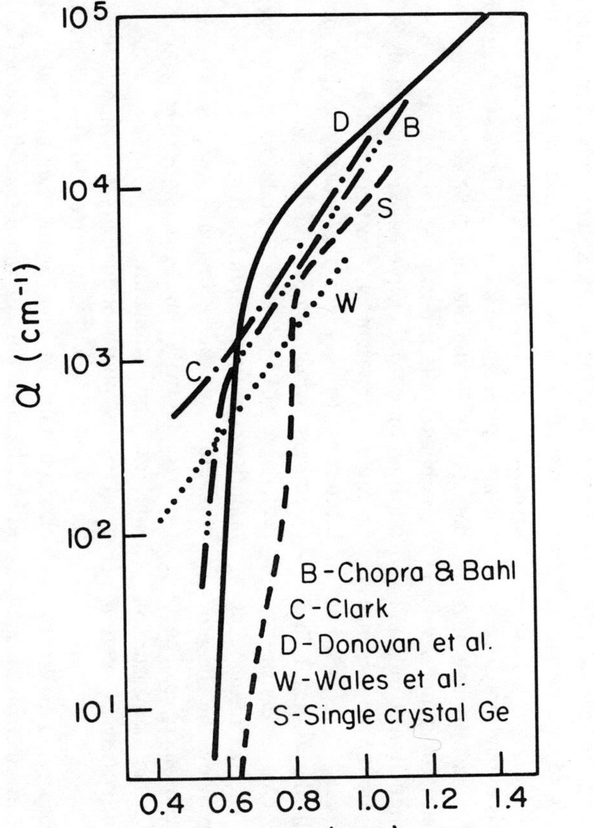

FIGURE 15. Comparison of the dependence of the absorption coefficient of amorphous Ge films on photon energy as obtained by several different workers. Results on single crystal Ge are also shown (Chopra and Bahl[183]).

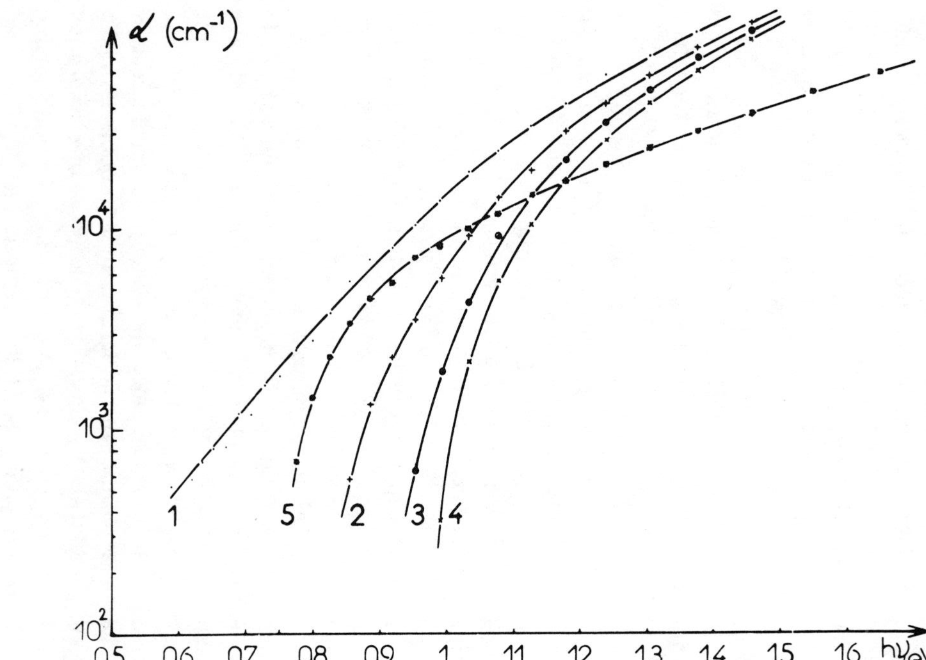

FIGURE 16. Absorption coefficient as a function of photon energy for a film of amorphous Ge: (1) as-deposited; (2) annealed at 200 C; (3) annealed at 300 C; (4) annealed at 400C; (5) crystallized (Theye[230]).

increasing with annealing temperature. As opposed to results of Theye[230] on amorphous Ge, the edge of amorphous Si remained lower than that of crystalline Si throughout the entire annealing range.

Beaglehole and Zavetova[229] measured the optical properties of a series of amorphous $Ge_x Si_{1-x}$ alloys and found a smooth variation of both the absorption edge and the absorption band with composition as x varies from 0 to 1. Thus no effects of compositional disorder appear.

To summarize, the position and shape of the absorption edges in both amorphous Ge and Si are very sensitive to the quality of the film. As-deposited films generally show an exponential band tail, but this tail is not intrinsic to the material and disappears upon annealing below the crystallization temperature. Since the EPR[136] and structural[154] studies show that such annealing results in a sharp reduction in void density, it can be concluded that the absorption tail arises from the presence of dangling bonds in the material. Furthermore, the extended nature of the absorption indicates that the states resulting from these dangling bonds do not have discrete energy values, as is generally the case for localized imperfections in crystalline semiconductors. The most likely reason for this is that the voids are not just single vacancies or divacancies, but represent roughly 15-30 missing atoms. Thus, not only are the voids structurally distinct, but the long-range disorder of the material already manifests itself over the volume of a void. In addition, the atomic positions can be expected to be particularly distorted on the surface of a void in amorphous films.

On the other hand, the voids extend only over a range of the order of 10 Å. Thus, models that interpret the band tails in terms of donor levels tailing down from the conduction band and acceptor levels tailing up from the valence band[147,183] cannot be correct.[241] This arises because such a situation will always result in a redistribution of electronic charge density to achieve an equilibrium state, the net result being the donor levels must end up below the acceptor levels. The difference between the energies of such localized states in the gap is, in fact, a measure of the correlation energy, U, discussed in section III.[123]

The observation of an absorption peak near 0.23 eV by Tauc et al.[236] in electrolytic and sputtered amorphous Ge shows that a well-defined level can exist in this material. This peak has also been observed in crystalline Ge that has been subjected to electron[242] or x-ray[243] bombardment, and can be associated with a Ge vacancy occupied by two electrons (when the level is full).[244] Thus, the results of Tauc et al.[236] indicate that electrolytic and to a lesser extent sputtered amorphous Ge films can have monovacancies as well as relatively large voids. Further evidence for this conclusion is the observation by Tauc et al.[236] of an intense absorption peak near 0.07 eV in electrolytic amorphous Ge. This absorption is very likely due to singly occupied Ge vacancies. The fact that the 0.23 eV peak is absent in the more ideally amorphous films of Donovan et al.[149,150] implies that monovacancies are not an essential feature of amorphous Ge, consistent with the structural model of Polk,[144] in which they do not appear.

Although Brodsky et al.[136] only measured the absorption of their amorphous Si films down to 0.5 eV photon energy, the absence of any structure below the edge is evidence against the existence of a large density of monovacancies in their films, since the doubly occupied acceptor level in Si should lead to an absorption in the 0.5 to 1.0 eV range.[245] Indeed, a strong absorption near 0.7 eV appears in radiation-damaged crystalline Si at low temperatures.[246] However, the monovacancy in crystalline Si is unstable at room temperatures, annealing out near 140K, so that we should not expect a significant density in amorphous films at 300K. The important defects in crystalline Si at room temperature are the divacancy and vacancy-impurity pairs. However, the absorption peaks corresponding to such simple defects all lie below the 0.5 eV limit of the experiments of Brodsky et al.,[136] and so no information as to their presence in amorphous Si is available at present. In particular, it would be interesting to look for the vacancy-oxygen pair (A-center), which should produce an absorption peak near 0.17 eV, as a guide to the oxygen impurity concentration in the films.

There have been few attempts to measure photoconductivity in amorphous Ge and Si. Konorov and Romanov[168] were not able to observe any photoconductivity in evaporated Ge films and Clark[147] reported only miniscule photocurrents in amorphous Ge, even at 4.2K where the dark current was essentially zero. Clark also failed to detect any evidence for thermally stimulated currents at 77K. On the other hand, Chittick et al.[134] observed an increase in conduction in glow-discharge-deposited amorphous Si of up to a factor of 10^3 upon low-level (100 ft candles) illumination. It should be recalled that the dark conductivity of these films was extremely small.

Photoemission experiments have been carried out on both amorphous Ge[228,247] and amorphous Si.[248] Donovan and Spicer[228,247] measured photoemission from both crystalline Ge and amorphous Ge films deposited in high vacuum. By cesiating the surface, photoelectron energy distributions were obtained for photon energies between 2 eV and 11.8 eV. In the amorphous films they observed a single broad maximum, which moved to higher energies with increasing photon energy in increments equal to the increment in photon energy. This is strikingly different from the results on crystalline Ge, and implies that **k**-conservation is not an important selection rule for optical transitions in amorphous Ge and also that the conduction band is without significant structure.[249] Annealing of the amorphous film above 400C yielded photoemission results which recaptured the structure of the crystalline material.

Although there was a major difference in the forms of the photoelectron energy distribution at different photon energies, the relative sharpness of the leading edge was quite similar. This is not what would be expected from the models discussed in section III, in which it was predicted that amorphous materials should exhibit valence band tails due to their disorder. The lack of **k**-conservation in optical transitions is no surprise, but the observation of a structureless conduction band is difficult to explain. With the assumption that optical matrix elements are independent of energy,[249] Donovan and Spicer[247] derived an optical density of states for amorphous Ge in which the valence band exhibited only a single broad peak about 1.5 eV below its edge. The structure of the crystalline valence band was virtually absent and the states were shifted to higher energy in the amorphous material. As will be discussed in sub-section B, the assumption of energy-independent matrix elements is somewhat doubtful.

Peterson et al.[248] measured photoemission from amorphous Si films deposited at 77K, and also obtained a single broad peak that increased with increasing photon energy. However, as opposed to the amorphous Ge results of Donovan and Spicer,[247] Peterson et al.[248] found distinct evidence for a valence band tail extending up to the Fermi energy of amorphous Si. Although the photoelectric yield decreased after a mild anneal, the band tail largely remained. It is unlikely that this represents a real difference between amorphous Ge and Si, which are in all other respects quite analogous. More likely, the band tail in the work of Peterson et al.[248] reflects the increased void density obtained from deposition at 77K. If so, further annealing below the crystallization temperature should greatly reduce the tailing. This experiment has not yet been done.

Piller et al.[250] measured electroreflectance in a series of Ge films deposited on substrates at different temperatures and thereby running the gamut from course-grained polycrystalline to amorphous. In crystalline Ge, five major static electroreflectance responses can be identified: E_o (0.6 - 0.8 eV), associated with the absorption edge; $E_o + \Delta_o$ (0.9 - 1.1 eV), associated with the lower spin-orbit-split valence band; E_1 (2.1 - 2.2 eV), associated with a critical point along the (111) direction; $E_1 + \Delta_1$ (2.3 - 2.4 eV), the spin-orbit-split companion to E_1; and E_2 (4.3 - 4.6 eV), associated with several higher-energy critical points. Piller et al.[248] found that the size of the electroreflectance signal monotonically decreased with decreasing substrate temperature and thus increasing disorder throughout the polycrystalline range for all five responses. The spectral positions remained constant in this range. However, upon the polycrystalline-amorphous transition, the signals associated with E_1, $E_1 + \Delta_1$ and E_2 suddenly disappeared entirely, while the peak energies of the E_o and $E_o + \Delta_o$ responses dropped about 0.1 eV energy. Thus, the only critical points that are preserved in the amorphous state are those associated with the fundamental edge. Although they were considerably weakened relative to the crystalline response, it is still difficult to explain why they should remain at all. It is even more surprising that some evidence of the spin-orbit-split valence band remains in the amorphous state, especially in view of the fact that there has never been any sign of such structure in the optical absorption.[150] Electroreflectance is ordinarily a particularly sensitive method for obtaining the energy gap of semiconductors, and the results on amorphous Ge, if real, indicate a decrease of the gap relative to the crystal.

Far infrared absorption on relatively thick evaporated, sputtered, and electrolytic films of amorphous Ge has been measured by Tauc et al.[234] The sputtered films, known to contain about 5% oxygen, exhibited a strong absorption peak near 0.09 eV. Since this is just the position of the dominant reststrahlen absorption in tetragonal GeO_2 [252] and since such a peak has been observed in oxygen-contaminated crystalline Ge that has been subjected to intense irradiation,[253] it is reasonable to assume that this peak is due to the oxygen impurities. The strong absorption near 0.07 eV in the electrolytically deposited films, also seen in the sputtered films, is most likely due to singly ionized monovacancies, as previously suggested. Finally, there is a weak absorption near 0.03 eV in electrolytic amorphous Ge, right in the optical-phonon range of crystalline Ge.[254] In crystalline Ge, the optical phonons are infrared-inactive, but Zallen[255] has shown that this is only a consequence of the fact that the diamond structure contains less than 3 atoms per primitive cell. Since this primitive cell structure is absent in amorphous Ge, a reststrahlen absorption becomes possible. Furthermore, since weak one-phonon peaks have indeed been observed in neutron-irradiated diamond[256] and Si,[257] materials with the same structure as Ge, it is likely that the 0.03 eV peak is just a disorder-induced reststrahlen absorption. Further evidence that this is the case comes from Raman spectroscopy,[258] in which a peak near 0.03 eV is also obtained.

Smith et al.[258] investigated the Raman spectra of several types of amorphous Si and Ge films. First-order Stokes peaks were obtained in amorphous Si near 0.02 eV and 0.06 eV; second-order Stokes and first-order anti-Stokes scattering were also observed. Shuker and Gammon[259] recently showed that breakdown of k-conservation in amorphous materials leads to a simple method for obtaining the density of phonon modes from the first-order Stokes spectrum. Smith et al.[258] attempted to estimate the phonon spectrum of amorphous Si by applying a Gaussian broadening to the calculated spectrum of crystalline Si.[260] The two major peaks obtained then agree with those found from the Raman scattering, and the general shape of the calculated phonon spectrum resembles that estimated by reducing the Raman spectrum in the manner suggested by the derivation of Shuker and Gammon.[259] When Smith et al.[258] crystallized their films by heating, they obtained the sharp Stokes lines observed in crystalline Si.[261] The Raman spectrum of amorphous Si was found to be insensitive to annealing. The fact that the phonon spectra of crystalline and amorphous Si strongly resemble each other implies that primarily near neighbors contribute to the vibrational modes. The absence of a high-frequency tail to the spectrum of amorphous Si is evidence against a significant density of localized phonon modes, even in the as-deposited films.

Very different results were obtained by Beserman et al.,[262] who studied the Raman spectra of evaporated amorphous Si films and compared them to the corresponding spectra of polycrystalline and single crystalline Si. Single crystal Si exhibits a sharp peak near 0.06 eV. Beserman et al. found that this peak was shifted up in energy by less than 1% and broadened only slightly in the polycrystalline films, but in the amorphous films it did not appear at all. Instead a new broad band about 20 times the width appeared near 0.08 eV. On the other hand, all of the other sharp peaks of the polycrystalline films were preserved in the amorphous films. It is not clear what these results indicate or why they differ from those of Smith et al.[258] The broad band could be a manifestation of phonon modes associated with the surfaces of voids in the film, but it was not investigated as a function of annealing.

4. *Band Structure*

The first attempt at estimating the density of electronic states of amorphous Ge was made by Herman and Van Dyke.[263,264] Stimulated by the photoemission results of Donovan and Spicer[247] and misled by the density determinations of Clark,[147] Herman and Van Dyke assumed that the density of states of amorphous Ge could be obtained directly from calculations of the crystalline density of states,[265] but with a larger interatomic spacing chosen to match the 28% density deficiency by Clark. As previously indicated, we now know that this density deficiency is almost entirely due to microscopic voids in the film and that the nearest neighbor separation in amorphous and crystalline Ge is essentially the same. In any event, the model of Herman and Van Dyke leads to semimetallic conduction, certainly not the case for amorphous Ge. An interesting feature of the calculation, however, is that under the assumptions of constant optical matrix elements and no k-conservation, ϵ_2 for dilated crystalline Ge exhibits a peak at 3.1 eV, considerably lower in energy than the major, 4.5 eV peak of real crystalline Ge and fairly near the observed peak at 2.7 eV in amorphous Ge.[247] In addition to the obvious defects of the model of Herman and Van Dyke, it is also clear that the bands obtained from dilated crystalline Ge represent an unstable state which should immediately relax back to the minimum energy equilibrium state with an enormous restoring force.

A more sophisticated attempt to calculate the band structure of amorphous Ge was performed by Brust,[266,267] who used a pseudopotential approach in which the atomic form factor was chosen to be identical to that of crystalline Ge. This acknowledges the equivalent near-neighbor environments of the two materials. Next, the pseudopotential itself was reduced by 9% to take into account the average density deficiency in amorphous Ge. The band structure was then calculated neglecting the disorder.[266] Finally, Brust assumed the long-range disorder could be accounted for by artificially introducing a large phonon scattering term, independent of energy and momentum. With these assumptions, ϵ_2 was calculated. Adjusting the magnitude of the electron-phonon scattering matrix element, an excellent fit was obtained to the observed curve.[149] Unfortunately, the position of the peak in ϵ_2 was found to be a sensitive function of the scattering matrix element. With the best fit for ϵ_2, the absorption edge turned out to be 0.3 eV, lower than any of the observed values. Brust[267] also estimated photoemission energy-distribution curves by assuming that **k** is a conserved quantity to within the uncertainty introduced by the large phonon scattering and modifying the calculated optical transition with a free-electron escape function. Brust suggested the results thus obtained were in good agreement with the experimental curves.[247] However, Spicer and Donovan[268,269] have pointed out that the agreement is, in fact, not very good as far as accounting for the independence of the shape of the leading edges of the energy-distribution curves with varying photon energy. This latter result is evidence that **k** is not all conserved in the transitions yielding photoemitted electrons. Spicer and Donovan[269] concluded that there is no experimental evidence that any features of the crystalline band structure are preserved in the amorphous state, as opposed to the predictions of Brust's calculation. However, the major objection to Brust's model is not so much the disagreement with the photoemission results but the fundamental fact that it is based on a microcrystallite model, and this is not consistent with the structural studies described previously.

Chakraverty[270] examined a microcrystallite model for elemental amorphous semiconductors, based on the tight-binding approximation of Leman and Friedel.[271] Chakraverty assumed that amorphous Ge and Si consisted of small 10 Å microcrystallites separated by totally disordered interfaces, and used a virtual-crystal approximation for the crystal potential. In this way, and using the result that the potential of an atom located in a disordered interface must be larger than that of an atom within a microcrystallite, Chakraverty was easily able to show that the effective energy gap always decreases in the disordered structure. This red shift of the gap upon disordering is in agreement with the explicit calculation of Brust.[267]

Stern[272] examined the band tail of amorphous Si by assuming that the electronic potential was that of the crystal modified by a Gaussian fluctuation, similar to the impurity model of Halperin and Lax.[75,78] The perturbing potential is then characterized by an rms amplitude, ΔV, and a correlation length, L. The latter was chosen to be 6 Å, the distance over which the short-range order of the crystal appears to be preserved.[154] The results then obtained for the band tail were interpolated to match the density of states deep in the band, analogous to a previous calculation of Kane.[73,273] The optical absorption coefficient, α, was then calculated, using $\mathbf{k}\cdot\mathbf{p}$ estimates for the momentum matrix elements[274] and using modified wave functions to break the **k**-conservation selection rule. It was also necessary to reduce the wave-vector separation of the valence and conduction band edges by 40%. ΔV was then chosen equal to 0.89 eV, the value which gave the best fit to the observed optical absorption of amorphous Si. The fit obtained is quite good between 1.0 eV and 2.0 eV, but overestimates the absorption at low energies. The latter defect is most likely another manifestation of the spatial separation of the states in the valence and conduction band tails, as discussed in section III and in some detail recently by Fritzsche.[275,276]

Stern[272] also investigated the location of the mobility edge in amorphous Si by introducing an effective "bandwidth" for states in the tails. This function depends on the density of localized states at a particular energy, and thus decreases with depth into the tail. Using the value of ΔV estimated from the optical-absorption fit and the Anderson[87] criterion for localization, equation (20), Stern estimated that the mobility edges are located 0.11 eV above the crystalline valence band edge and 0.20 eV below the crystalline conduction band edge. In each case, the density of states at the mobility edge turned out to be approximately 10^{21} cm^{-3} eV^{-1}. This calculation should not be taken too seriously, since the Anderson criterion describes the critical ratio of disorder to bandwidth at which *all* states become localized, and thus is not strictly applicable to the mobility-edge problem. It would appear that the approximation of Stern underestimates the position of the mobility edge relative to the crystalline band edges, since the Anderson criterion is too strong for band-tail localization. Larger values for the separation of the mobility edge from the crystalline band edge have been obtained using semi-classical percolation techniques.[114,277]

Stern[272] estimated a mobility gap of 0.81 eV, with the Fermi energy twice as far from the valence band as from the conduction band. This leads to the prediction that amorphous Si is n-type, and comes about because the valence band tail is more spread out than the conduction band tail in Stern's model. The density of states at the Fermi energy was estimated as about 4×10^{20} cm^{-3} eV^{-1}, the majority of these tailing from the conduction band. These results are basically in agreement with conclusions drawn from the tunneling experiments of Sauvage et al.[195] Stern also calculated an intrinsic activation energy of 0.25 eV and a mobility of about 5 cm^2/V-sec beyond the mobility edges. These values are very difficult to compare with experiment, since the experimental results are at present rather contradictory. Furthermore, Stern did not attempt to consider the effects of either voids or hopping conduction in his model. Nevertheless, the calculation lends some quantitative support to the applicability of the CFO model even in a relatively ordered material like amorphous Si.

5. Conclusions

There has indeed been a flurry of investigation into the properties of amorphous Ge and Si in the past two years, and it is worthwhile to ask whether a clearer picture has emerged from the combined results of the more than 50 experimental papers that have been published since the review of Clark.[278] The answer is weakly affirmative – the picture is clearer, but far from transparent. The tentative conclusions that can be drawn at the present time are summarized below.

1. Perhaps the most important lesson we have learned is that results are extremely sensitive to the quality of the films being investigated. Both the quality and the purity are strongly dependent on the exact method used to prepare the films. Comparing the experimental results on, for example, glow-discharge-produced Si and evaporated Si is in most respects like comparing lettuce and cabbage.[134] In addition, it should be realized that the simple process of attaching electrodes to the material and applying a small electric field, as is usual in performing transport experiments, is likely to crystallize some of the material and completely invalidate the results.[217,279,280]

2. It now also seems clear from the structural work of Moss and Graczyk,[154] of Polk,[144] and of Brodsky et al.[136] that an ideal amorphous state does exist for amorphous Ge and Si, and that it is of a random-network nature. The ideal state is generally not obtained when the films are initially made, but there is evidence that it can be approached by annealing below the crystallization temperature.

3. The as-deposited films generally contain a large density of voids of the order of 5 Å in extent (Swiss cheese model), thus resulting in density deficiencies of 5 to 30% relative to the ideal amorphous state. These voids lead directly to a sufficient density of dangling bonds in the material to produce extrinsic conduction in the vicinity of room temperature. By annealing with crystlline Ge and Si, we can conclude that the predominant conductivity resulting from these dangling bonds is p-type.

4. At high temperatures when most of the voids have been annealed away, or in compensated films, intrinsic conduction appears with a much larger activation energy. This conductivity seems to be n-type (although there is some contradictory evidence), indicating a relatively larger electron mobility than hole mobility.

5. Hopping conduction exists, but it does not always predominate at room temperature. The drift mobility experiments of LeComber and Spear[191] indicate that hopping predominates in their glow-discharge-produced films below 240K, and that hopping takes place primarily in states in the band tail about 0.2 eV below the mobility edge, rather than in the vicinity of the Fermi energy. On the other hand, a large number of as-deposited films exhibit a $T^{1/4}$ -law dependence of the conductivity, as suggested by equation (28), up to room temperature and even above. This is evidence for the predominance of hopping conduction in the vicinity of the Fermi energy. Further support for this conclusion comes from the tunneling results. Since the films very likely contain large void densities and thus a sizeable acceptor-like "impurity" band, the primary conduction mechanism could be ordinary impurity conduction, which continues to dominate up to room temperature because of the relatively low mobility of holes below the valence-band mobility edge. As the films are annealed, the density of impurity states decreases, resulting in sharp decreases in conductivity. Eventually, intrinsic band-like conduction could dominate the hopping conduction as the ideal amorphous state is approached.

6. Near-ideal amorphous films exhibit sharp absorption edges,[150] essentially just as sharp as in the corresponding crystals. However, as discussed previously, this cannot be taken as evidence against the existence of a significant density of localized states in the mobility gap. The optical gap in amorphous Ge appears to be between 0.6 eV and 1.0 eV, depending on the method of preparation. The corresponding range for amorphous Si is even wider, 0.8 to 1.6 eV.[248]

7. The van Hove singularitites *within* the bands appear to have disappeared in the amorphous materials, in accordance with theoretical expectations. In fact, only a single broad peak remains in ϵ_2 in amorphous Ge and Si, and it appears to be connected with the short-range order; i.e. it results directly from the tetrahedrally oriented covalent bonding of these Group IV semiconductors. If this is the case and the peak in ϵ_2 is just a broadened version of the L-L interband transitions of the corresponding crystal, it should have the same pressure coefficient as that particular crystalline transition (7×10^{-6} eV/bar for Ge[239]). This has not yet been experimentally investigated. As previously discussed, the band-edge van Hove singularities appear to be preserved in the amorphous state; this is most likely due to the existence of mobility edges and the spatial separation of localized states tailing from different bands, rather than to the absence of states in the gap.

8. There does not appear to be much optical evidence for the peak near mid-gap proposed by Davis and Mott.[130] An absorption peak near 0.23 eV has been observed in electrolytic and sputtered films of amorphous Ge,[232,236] but this probably represents an acceptor band due to monovacancies.

9. The observation of well-defined activation energies, particularly at high temperatures and in annealed films, is evidence for the existence of sharp mobility edges. However, the behavior of the density of states is unknown in this region, so that it is still possible that these activation energies represent the effects of ordinary crystalline-type band edges. On the other hand, the absence of interior van Hove singularities is a strong point in favor of the mobility-edge interpretation.

10.Somewhat surprisingly, there is no evidence for significant effects of compositional disorder in amorphous Si-Ge alloys.[229] This is evidence against the applicability of the CFO model to amorphous Ge and Si, since, in this model, the additional disorder of the alloys should increase the extent of the band tails, and thus $g(E_f)$. On the other hand, if $g(E_f)$ primarily results from the void density, the additional disorder would not be expected to produce large effects, in accordance with the observations.

B. Se

Because of its commercial importance,[281-283] amorphous Se has been investigated extensively, and is one of the most understood amorphous semiconductors. As opposed to Ge and Si, Se is a Group VI element and requires only two nearest neighbors to fulfill its bonding requirements. One advantage of Se over Ge and Si is that amorphous Se can be quenched from the liquid state; i.e. bulk glassy samples can be made. A complicating feature is that Se crystallizes in two different structures.[284] The stable form below the melting point, 493K, is trigonal, the atoms being arranged in spiral chains so that the projection on a plane perpendicular to the axis is an equilateral triangle. The intrachain bonding is covalent, while the interchain bonding is due to relatively weaker Van der Waals attractions. The structure is obviously highly anisotropic. Two very similar metastable monoclinic forms of Se exist, α-monoclinic Se[285] and β-monoclinic Se[286]. The principle structural unit in both structures is not infinite Se chains as in trigonal Se, but finite molecules in the form of puckered Se_8 rings. Thus monoclinic Se is molecular (or monomeric) rather than polymeric. The intramolecular bonding is covalent, the intermolecular forces are Ven der Waals. In all three forms of crystalline Se, the covalent bond length is essentially the same, approximately 2.32 Å. In addition, the bond angles, 105° in the trigonal form, are only slightly distorted, up to $\pm 4,^\circ$ in the monoclinic forms. The closest Van der Waals separations in the three structures are also essentially equal, about 3.5 Å. Se is intermediate between S and Te in the periodic table, and indeed monoclinic Se is structurally very similar to orthorhombic S (which contains molecular S_8 rings), while trigonal Se has the same structure as crystalline Te.

1. Structure

From the knowledge that the short-range order of corresponding crystalline and amorphous materials is generally the same, it is not too difficult to infer models for the structure of amorphous Se from the existing crystalline allotropes. Since two types of short-range order can exist, polymeric chains as in trigonal Se and molecular Se_8 rings as in monoclinic Se, it is reasonable to assume that both can be simultaneously present in amorphous Se. Indeed, suggestions along these lines came relatively early and often,[287-294] and have been more or less confirmed by later work. Because of the covalent bonding requirements of Se and the similarity in nearest-neighbor distance in all the crystalline structures, we should always expect a first-neighbor coordination number of 2 and a nearest-neighbor distance of about 2.32 Å in any form of Se. Thus it is difficult to differentiate rings from chains by ordinary analysis of the x-ray radial distribution functions. Such radial distribution functions have been obtained by several groups,[295-301] and have been interpreted in very different manners: Hendus[295] postulated primarily polymeric chains; Krebs and Schultze-Gebhardt[296] suggested the presence of eight-membered and higher rings; Grimminger[297] assumed the predominance of Se_6 rings at 77K but chains at 300K; Andrievskii[298] concluded that primarily Se_8 rings exist at 290K, while short chains predominate above 340K; Henninger et al.[299] suggested that amorphous Se was essentially a random distribution of chains; Kaplow et al.[300] concluded that the structure was 95% Se_8 rings; and Breitling and Richter[301] ruled out Se_8 and Se_6 rings entirely, deciding in favor of randomly arranged chains. The agreement is less than spectacular.

An alternative approach is to study other physical properties such as viscosity above the glass transition temperature. Because the forces between Se atoms in different rings or chains are of the relatively weak Van der Waals type, the glass transition temperature of amorphous Se is significantly lower than those of amorphous Ge and Si. Values differ somewhat, depending on the type of measurement - elastic moduli[302] and dilatometric[303] data give 305K, while DTA experiments[304] suggest 315K. The high viscosity of amorphous Se between the glass transition temperature and the melting temperature, 490K, can be looked at as evidence for long chains.[290] Studies of liquid Se[289] indicate a 60% polymer concentration at the melting point, and polymerization theory[305] leads to the conclusion[292,303] that the average chain length is of the order of 2000 atoms. Assuming chains are terminated by dangling bonds, Massen et al.[306] inferred an average chain length of 10^5 atoms from susceptibility measurements, but this is probably an overestimate since it overlooks chain termination by, for example, halogen impurities.

A clearer picture of the structure of amorphous Se emerges from the infrared[294,307-312] and Raman[309-313] spectroscopy experiments. The infrared transmission of amorphous and α-monoclinic Se between 50 cm^{-1} and 300 cm^{-1} are compared in Figure 17,[310] and the resemblance is striking. The fundamental absorption of the Se_8 molecule at 95 cm^{-1} and 254 cm^{-1} are both found as strong peaks in the absorption of amorphous Se, while the Se_8 absorption at 120 cm^{-1} appears as a shoulder in the amorphous spectrum. In addition, most of the weaker phonon-combination bands in the 250-800 cm^{-1} range can be correlated with transitions of the Se_8 rings in α-monoclinic Se.[309,310] This is strong evidence for the presence of Se_8 molecules in amorphous Se. The only sharp absorption peak of amorphous Se that is absent in α-monoclinic Se is near 135 cm^{-1}. This is fairly close to the

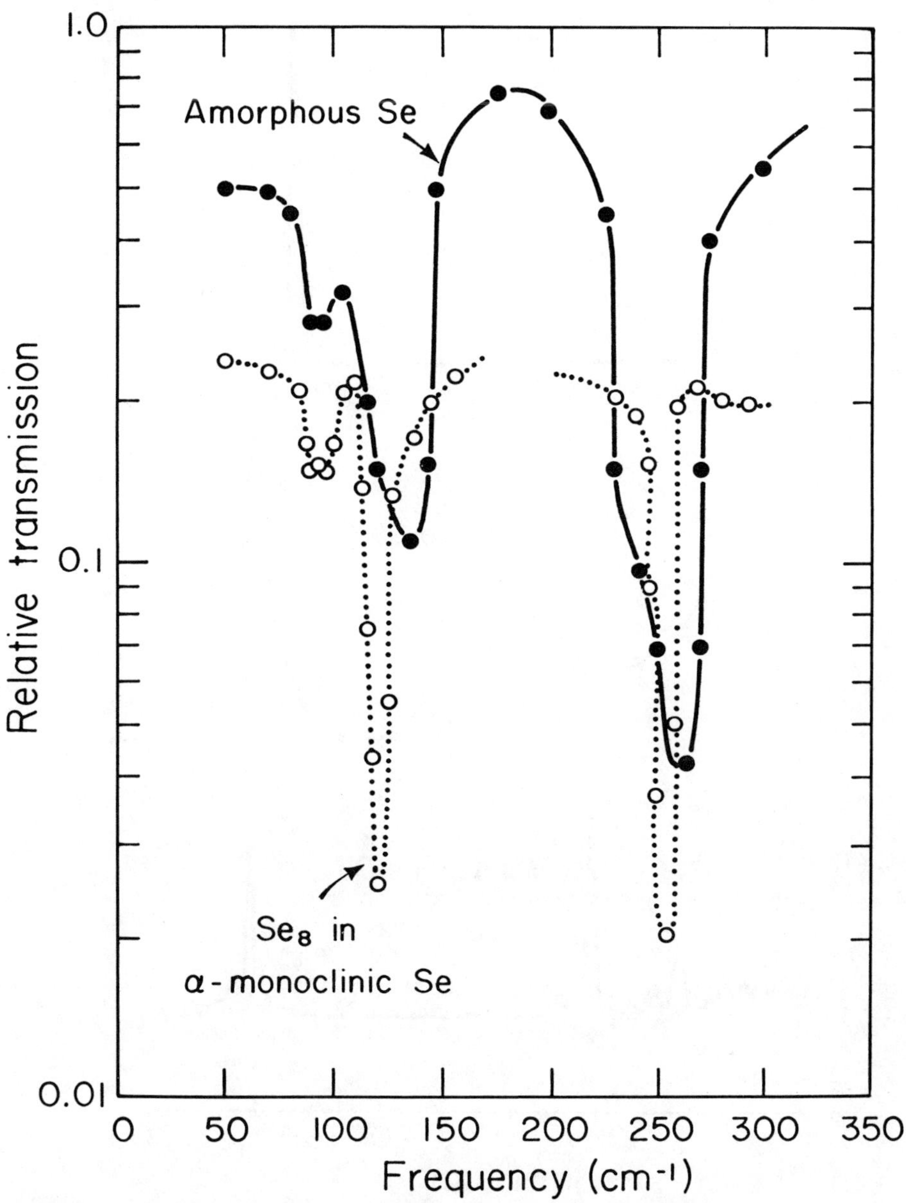

FIGURE 17. Comparison of the infrared spectrum of amorphous Se with that of Se_8 in α-monoclinic Se (Zallen and Lucovsky[312]).

fundamental absorption at 144 cm^{-1} in trigonal Se and consequently has been interpreted as a vibrational band of Se chains in amorphous Se,[310] although this is not as clear as the other assignments. The vibrational bands are all somewhat broadened in the amorphous material, as might be expected.

The Raman spectra of trigonal, amorphous, and α-monoclinic Se at 4K are shown in Figure 18.[313] The major peaks of amorphous Se, near 250 cm^{-1} and 110 cm^{-1} are coincident with the dominant peaks in α-monoclinic Se, except for a broadening in the amorphous phase. Similarly, even the weaker peaks of amorphous Se, near 50 cm^{-1} and 80 cm^{-1}, appear in the α-monoclinic spectrum. In addition, two shoulders exist in the amorphous material, near 140 cm^{-1} and 235 cm^{-1}, very close to the fundamental modes of trigonal Se at 143 cm^{-1} and 237 cm^{-1}. These shoulders in the amorphous Raman spectrum also appear in the infrared absorption and most likely represent chain-like vibrations. On the other hand, there is

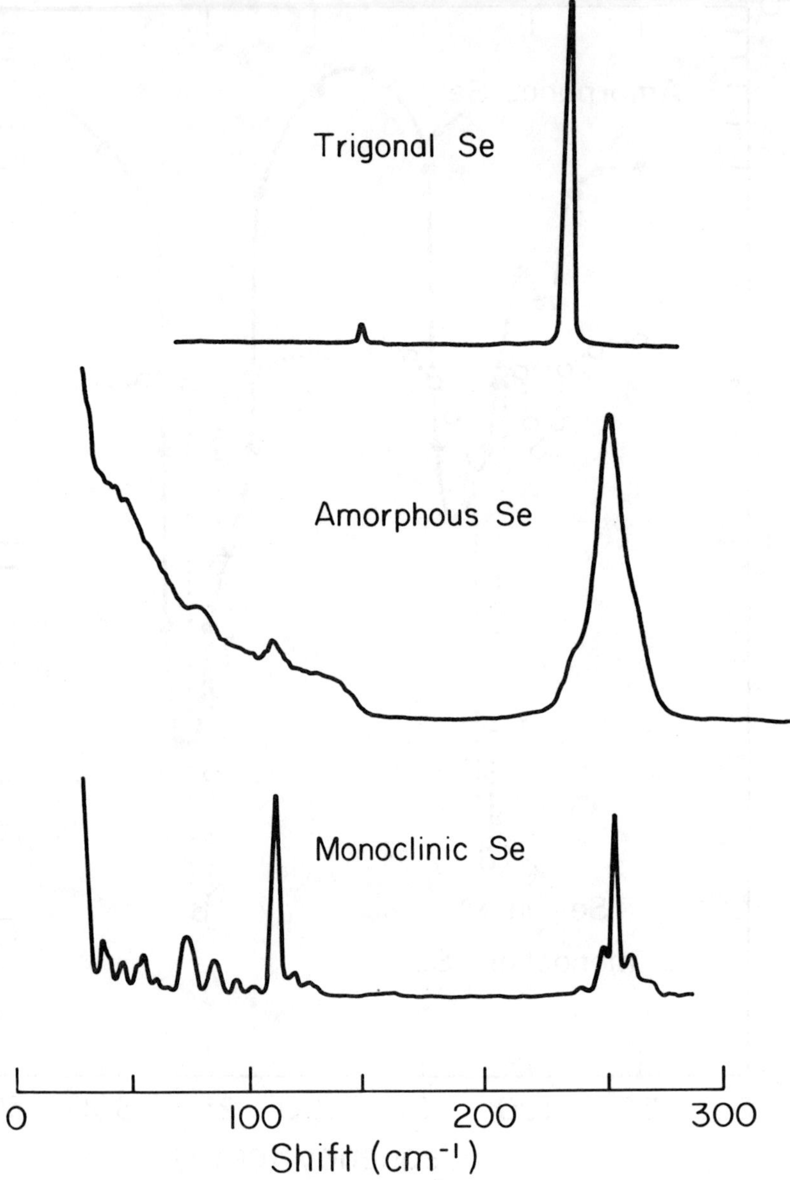

FIGURE 18. Comparison of the Raman spectra of trigonal, α-monoclinic, and amorphous Se (Zallen and Lucovsky[312]).

some structure in the monoclinic Raman spectrum at these energies, and the presence of polymeric chains cannot be verified with certainty from these optical experiments. On the other hand, there is no doubt that Se_8 rings are present in amorphous Se.

At present, the model of amorphous Se as combinations of molecular Se_8 rings and polymeric chains is consistent with all the structural studies. The infrared and Raman experiments confirm the existence of rings, while the viscosity[314,315] and copolymerization[316,317] experiments provide the best evidence for chain-like units. The ratio of rings to chains and the average chain length are still in doubt, and may depend strongly on the method of preparation.

2. *Electrical Properties*

The dc conductivity of amorphous Se has been studied often over the years,[318-327] but no consistent picture has yet arisen. This is primarily because of the extreme insulating nature of the pure material. Room-temperature conductivity values of $10^{-11}\ \Omega^{-1}\ cm^{-1}$ to $10^{-17}\ \Omega^{-1}\ cm^{-1}$ have been reported, although the lower figures are most likely representative of intrinsic Se. Lacourse et al.[327] have shown that only 50 ppm oxygen impurities lead to a sharp increase in conductivity from $10^{-17}\ \Omega^{-1}\ cm^{-1}$ to greater than $10^{-11}\ \Omega^{-1}\ cm^{-1}$. The dc conductivity is thermally activated, with an activation energy in the range 0.95 to 1.10 eV.[323,326,328] The higher values are associated with the lower conductivity material.[323] An interesting result is that the conductivity of amorphous Se merges with that of liquid Se at the melting point without discontinuity in either conductivity or activation energy.[320,323,329-331]

A complication in measuring the dc conductivity is the non-linear I-V characteristics that appear at relatively low fields, 10^2 V/cm.[324] The non-ohmic region is characteristic of space-charge-limited conduction. Lanyon[324] found a $V^{3.8}$ dependence of current on voltage in several films, a relationship expected from an exponential distribution of traps in the gap.[332] Analysis of the space-charge-limited currents indicate a density of localized states near the valence-band mobility edge of the order of $10^{20}\ cm^{-3}\ eV^{-1}$.[323,324]

No Hall data have been reported for amorphous Se, since the extremely low conductivity makes the Hall voltage undetectable. Gobrecht et al.[333,334] measured the Hall effect in liquid Se, and found a p-type Hall mobility, which increased from 20 cm^2/V-sec to 50 cm^2/V-sec between 550K and 700K. A maximum appeared near 700K, and the Hall mobility decreased exponentially above that point with a behavior characteristic of optical-phonon scattering. It is not clear from this data whether the maximum indicates just a change in the predominant scattering mechanism or perhaps a fundamental change from hopping to band-like conduction.

Thermoelectric-power measurements have been carried out on both amorphous[335] and liquid[335-338] Se. Henkels and Maczuk[337] observed a p-type thermopower in liquid Se from 500K to 800K and concluded that conduction was intrinsic, that the energy gap was 2.3 eV, and that the hole mobility decreased with increasing temperature proportional to $T^{-3/2}$. Assuming an effective mass equal to the free-electron mass, they estimated a value of about 500 cm^2/V-sec for the conductivity mobility at 700K. This is an order of magnitude larger than the observed Hall mobility, but an effective mass of just 5 times the free-electron mass would put the two values in agreement. Abdullaev et al.[338] found that the thermopower of ultra-pure (1 ppm impurities or less) liquid Se was p-type but just slightly contaminated (10 ppm impurities) Se exhibited an n-type thermopower. On the other hand, ultra-pure Se that had been deoxygenated was n-type, indicating that even a very slight concentration of oxygen impurities introduced acceptor levels. These acceptors are quickly compensated by other contaminants. This sensitivity to small concentrations of impurities encourages a cautious approach to interpreting the thermopower data. Dutchak et al.[335] measured the thermoelectric power of amorphous and liquid Se contaminated with about 10 ppm impurities from 300K to 700K. In the amorphous phase, the thermopower was n-type and relatively small in magnitude ($\sim 100\ \mu$V/K). The magnitude decreased with increasing temperature until the material crystallized near 370K, with a sharp sign reversal. At the melting point, the sign of the thermopower again reversed to n-type behavior, and the magnitude exponentially increased with increasing temperature. It is clear, primarily from the results of the drift-mobility studies which will be discussed shortly, that the n-type thermopower in the liquid phase is entirely due to impurities, so that the corresponding data on the amorphous phase should not be considered representative of pure Se. It is difficult to reconcile the extreme sensitivity of the thermopower to small impurity concentrations with the relative insensitivity of the electrical conductivity to most types of doping.[337,339] However, as previously discussed, Lacourse et al.[327] showed that oxygen impurities can indeed drastically affect the conductivity most likely by introducing acceptor levels that induce extrinsic p-type semiconduction.

Lakatos and Abkowitz[326] measured the ac conductivity of amorphous Se up to 6 x 10^{10} Hz and over a temperature range of 77 to 320K. The ac conductivity increased as ω^n up to 10^8 Hz at all temperatures, with n decreasing from 1.08 at 77K to 0.85 at 320K. Above 10^8 Hz, the ac conductivity appeared to saturate, while at low temperatures, the ac conductivity was essentially temperature-independent. The frequency dependence of the conduction is characteristic of a hopping process.[128,129] Application of Equation 29 to the data of Lakatos and Abkowitz predicts a density of states at the Fermi energy in excess of 10^{18} cm^{-3} eV^{-1}. This is unlikely to arise from a CFO-type band overlap, since amorphous Se is a relatively ordered material with a large mobility gap. It appears much more probable that the observed hopping conduction occurs via impurity levels rather than at the intrinsic Fermi energy. The impurities could easily account for a value of $g(E_f)$ of the order of 10^{18} cm^{-3} eV^{-1}.

Drift-mobility investigations of amorphous Se have been performed extensively.[323,340-350] The hole mobility consistently dominates the electron mobility throughout the temperature range investigated (200 to 300K).[323] The room-temperature hole mobility is 0.14 cm^2/V-sec, while the electron mobility is approximately 5 x 10^{-3} cm^2/V-sec. Both the electron and hole drift mobilities are thermally activated, with activation energies of 0.28 eV and 0.14 eV respectively, near room temperature.[323,341] More recently, the hole drift mobility has been reinvestigated,[344,348] and the activation energy appears to increase to 0.23 eV between 200K and 260K. These results are shown in Figure 19. Below 200K, the hole transit time is no longer a linear function of field, and a drift mobility cannot be defined.[348] The drift mobilities of both electrons and holes above 200K are essentially independent of pressure,[349] strong evidence against the predominance of hopping conduction in this temperature range. It is thus extremely likely that the drift mobility is trap controlled, as originally suggested by Spear.[341] It is possible that hopping conduction predominates below 200K, and that field-induced tunneling between localized states is responsible for the lack of a well-defined drift mobility in this region.[348,351]

The fact that the room-temperature drift mobility of holes is trap controlled implies that the conductivity mobility is significantly larger than 0.14 cm^2/V-sec and, in fact, could well approach the order of 200 cm^2/V-sec, the value obtained for the Hall mobility of liquid Se at 550K. Interestingly enough, this is also the order of the hole mobility in single-crystal trigonal Se.[352] Spear and Lanyon[342] estimate a hole mobility of 60 cm^2/V-sec for amorphous Se, also of the same order. The pressure independence of the drift mobility taken together with the sharp increase of conductivity with pressure[349] indicate a decrease of mobility gap with pressure for

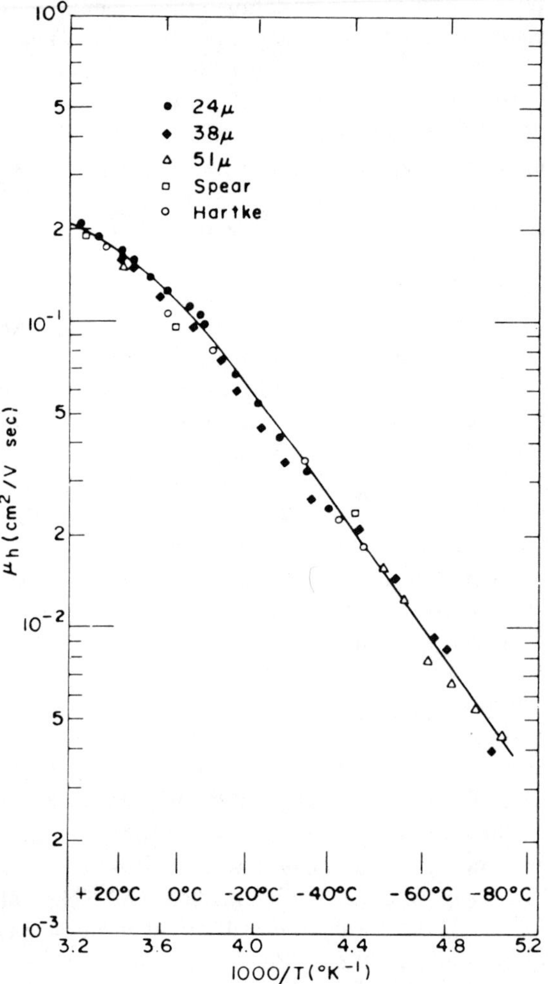

FIGURE 19. Temperature dependence of the hole drift mobility for amorphous Se films of three different thicknesses and values obtained by other workers. The straight-line portion at the lower temperatures gives an activation energy of 0.23 eV; the data for temperatures higher than -10C show distinct curvature (Tabak[348]).

amorphous Se. Once again, similar results apply to trigonal Se.[349,353,354]

To summarize, the conduction observed in both liquid and amorphous Se that exhibits an activation energy near 1.1 eV appears to be intrinsic. This conclusion is reinforced by the optical results discussed in the next subsection. This intrinsic conduction is p-type, due to the larger mobility of valence-band holes as compared to conduction-band electrons. The hole conduction is almost definitely band-like rather than of a hopping nature. Hopping among impurity levels has been observed in the ac conductivity. Se appears to be a material in which oxygen impurities introduce acceptor levels, but these acceptors are easily compensated by other type impurities. Consequently, intrinsic conduction is commonly observed.

3. *Optical Properties*

The ir and Raman spectra of amorphous Se have already been discussed in subsection 1, since they give information primarily about the structure of the material. In this subsection, we shall be concerned with electronic (visible and uv) transitions and photoconductivity.

Extensive optical experiments[281,307,355–377] have been performed in the vicinity of the absorption edge, and the results are remarkably consistent, as is shown in Figure 20. The edge occurs in the vicinity of 1.8 eV and is not very sharp, extending from about 1.5 eV to 2.2 eV. Siemsen and Fenton[377] measured the temperature dependence of the absorption edge from 77K to 700K, and found that the edge is relatively sharp (1.7 to 2.2 eV) at the lowest temperatures, but broadens strikingly with increasing temperature,

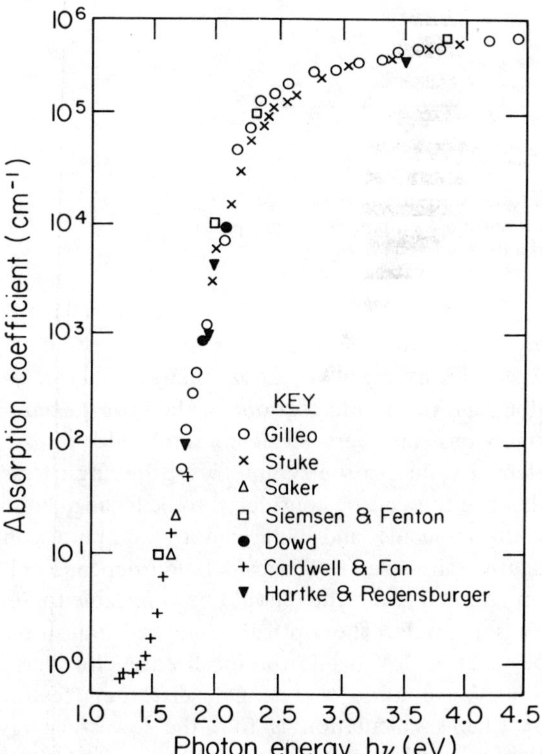

FIGURE 20. Several results for the optical absorption coefficient as a function of photon energy for amorphous Se, showing the reproducibility of the edge (Lucovsky and Tabak[281]).

finally extending over a 1.7 eV range (0.5 to 2.2 eV) in the liquid region. Absorption above 2.4 eV is essentially independent of temperature. Kandare[378] found that the reflectivity of amorphous Se remains high at least through 12 eV.

Although amorphous Se has been studied extensively for over a century, its photoconductivity, the major reason for its commercial importance at present, was not observed until relatively recently.[366,379–382] The photocurrent is generally small, but because of the extremely low dark conductivity, it is several orders of magnitude larger than the dark current. Even in the early work, it was evident that the photoconductivity edge is about 0.6 eV greater than the absorption edge.[366,383] The situation is as shown in Figure 21. Recent experiments have confirmed this unusual result.[343,345,346,384–388] Tabak and Warter[343] showed that the rate of photogeneration of free carriers depends on both temperature and applied field, also an unusual situation in elemental semiconductors. Clearly, in amorphous Se, photons can induce electronic transitions into localized states, and these transitions consequently do not lead to electrical conduction. Tabak and Warter[343] were able to observe what appears to be a modified high-field Poole-Frenkel effect[389] in the photoconductivity for electric fields in excess of 10^4 V/cm, and Tabak and Scharfe[388] employed a configuration in which they could observe a transition from an initial emission-limited photocurrent to a steady-state space-charge-limited photocurrent.

One possible explanation of the non-photoconductive absorption is that it represents transitions between localized states in the valence and conduction band tails, although for reasons discussed in section III this is extremely unlikely. Davis[390] suggested that these transitions are across the mobility gap, but do not conduct until the photogenerated electron-hole pairs are freed from their mutual Coulomb attraction. This excitonic breakup is supposed necessary in amorphous, as opposed to ordinary crystalline, material because of the low carrier mobility of the former. However, as pointed out previously, the hole mobility in amorphous Se is most likely of the order of 20 cm^2/V-sec, not very different from that of some ordinary crystalline semiconductors, such as GaP.[391] Lucovsky[392] suggested that the non-photoconductive absorption represents a Frenkel exciton,[393] which can never lead to

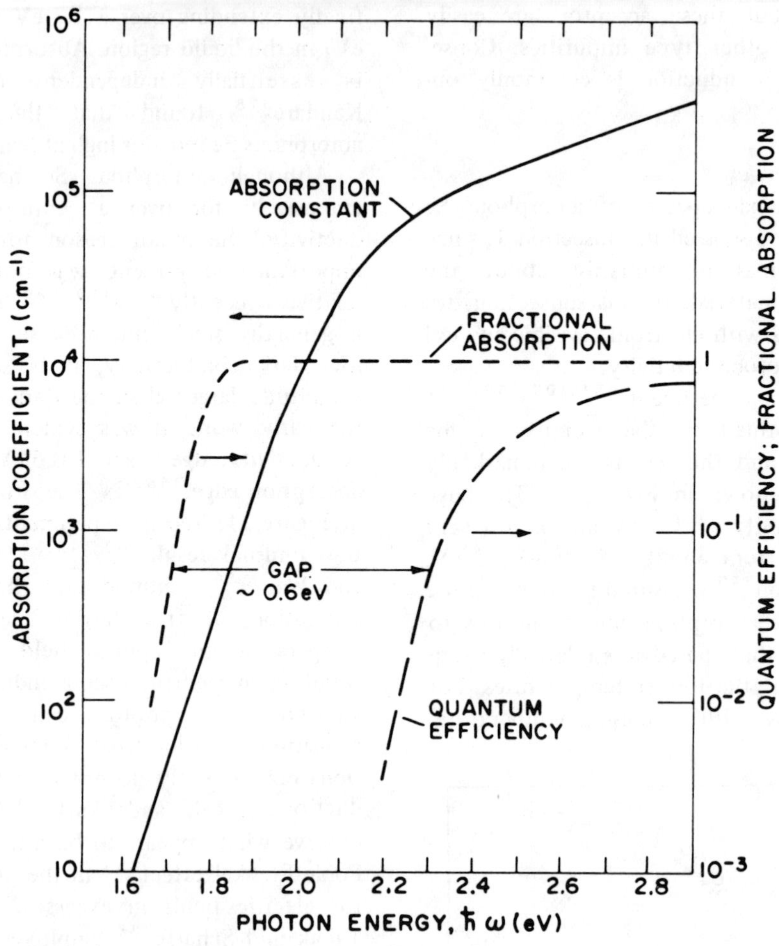

FIGURE 21. Dependence of the optical absorption coefficient and the high-fields quantum efficiency on photon energy for a 40 μ sample of amorphous Se. The fractional light absorption is shown as a dotted line (Lucovsky[392]).

conduction, since it cannot be ionized either thermally or by an applied electric field. With this model, Lucovsky was able to explain the temperature and electric-field dependence of the photogeneration rate[343] and the apparent frequency-dependent activation energy for thermal excitation of trapped carriers.[387] Lucovsky[392] further suggested that the Frenkel excitons represent molecular excitations of the Se_8 rings known to exist in amorphous Se.[310] There is a great deal of experimental support for this conclusion. It is clear from Figure 18 that the Raman spectrum of amorphous Se is dominated by vibrations of Se_8 rings. In addition, α-monoclinic Se also exhibits a non-photoconductive absorption edge,[394] and a similar result has been observed in orthorhombic S,[395,396] whose structural units are S_8 rings. Chen[397,398] has theoretically investigated the electronic energy levels of S_8 molecules by means of a molecular-orbital calculation, and has found that one of the lowest exciton states has only very small intermolecular transfer elements, due to the symmetry of the rings. Thus, this exciton can be considered to be localized on a given molecule and will lead to non-photoconductive absorption. The calculation depends only on symmetry, and thus should be applicable to Se_8 rings as well. The optically allowed transitions begin at 3.6 eV, while the localized exciton has a 3.1 eV excitation energy. One disturbing feature of Chen's calculation is that the lowest energy exciton, with a 2.2 eV excitation energy, is not localized. Although it is optically forbidden in a single molecule, so is the localized exciton. A possible resolution of this problem is that the shifts in energy of these lowest two excitations upon crystal formation reverse their order. An advantage of assuming that the major features of both the non-photoconductive and the photoconductive contributions to absorptions in amorphous Se are due to Se_8 ring excitations is that it accounts for the equal temperature coefficients of the two edges.

The electroreflectance of amorphous Se from 1.7 eV to 4.2 eV has been investigated by Weiser and Stuke,[399] who found only two broad peaks, near 2.1 eV and 2.2 eV. These values are near the photoconductivity edge, but are in the region where the absorption coefficient is already between 10^4 cm^{-1} and 10^5 cm^{-1}. This is very different behavior from trigonal Se, in which the electroreflectance signal is an order of magnitude weaker, is much sharper, and exhibits a great deal of structure in the 1.8 to 2.5 eV region.[400] The results for trigonal Se can be interpreted as transitions from the two spin-orbit valence bands to exciton levels just below the conduction-band edge. It would thus appear that the likely explanation of the electroreflectance peaks in amorphous Se is that they are due to excitonic absorptions of Se_8 rings, just as is the non-photoconductive component of the optical absorption. However, this cannot then account for the fact that the absorption coefficient exceeds 10^4 cm^{-1} at lower photon energies than 2.1 eV, nor does it explain the temperature independence of the electroreflectance signal, an unusual result for excitonic peaks. In view of the correlation between the 2.1 eV structure in the electroreflectance with the band gap estimated from both photoinjection and Faraday rotation experiments (as will be discussed subsequently), it is more reasonable to assume that the electroreflectance peaks represent somewhat washed out critical behavior in the vicinity of the density-of-states gap of amorphous Se. This is possible only if the band tails due to disorder are not very extensive in amorphous Se, but this could be the case if the electronic structure is dominated by transitions involving Se_8 rings.

Electroabsorption has also been studied in amorphous Se,[401-403] and a Franz-Keldysh-type[404,405] red shift of the absorption edge with electric field has been observed. The effect is about two orders of magnitude smaller than in crystalline Se.[401] This shift contributes a spurious peak near 1.8 eV in the electroreflectance signal at the first harmonic.[399]

Mort and Scher[406] recently measured the Faraday rotation in amorphous Se, an experiment first performed on the material by Becquerel[407] nearly one hundred years earlier, and sporadically reinvestigated since that time.[408,409] Mort and Scher also developed a theory for interband Faraday rotation in amorphous semiconductors, based on a calculation of the frequency dependence of the index of refraction. They assumed that disorder forces the partial breakdown of k-conservation over a fraction of the Brillouin Zone, analogous to the optical calculations of Tauc et al.,[222] and introduced a disorder parameter, z, related to the extent of the nonconservation of **k**. They then found that the best fit to their Faraday rotation data in the 1.3 to 1.9 eV range occured for z=0.2, a value that indicates amorphous Se exhibits about twice the disorder of amorphous Ge.[222,227] Mort and Scher estimated a band gap of 2.1 eV from their best fit. This value represents what would be the optical gap were it not for the excitonic absorptions of the Se_8 ring molecule. It could be somewhat smaller than the mobility gap, if significant band tails exist in amorphous Se. However, the same value for the band gap emerges from experiments on the photoinjection of carriers from metallic electrodes into the semiconductor,[350] and this figure should represent the actual mobility gap. We thus can conclude that band tailing is not a very significant feature of the band structure of amorphous Se. As previously indicated, the same 2.1 eV figure is obtained from the first broad electroreflectance peak.[399]

Amorphous Se also exhibits a photocrystallization effect,[410,411] which becomes extremely large when the photon energy exceeds 2.2 eV.[412] Thus, just as in photoconductivity and electroreflectance, the excitonic absorption is not very significant.

4. *Band Structure*

The only attempt at calculating the band structure of amorphous Se up to the present has been that of Kramer et al.,[413] who used a pseudopotential approach in which the disorder was handled by introducing a **k**-dependent lifetime broadening. Similar to the band calculation of Brust[266] for amorphous Ge, Kramer et al. assumed that the atomic form factor was identical to that of trigonal Se. This assumption is not as obviously valid for Se as it is for Ge, since the short-range order of amorphous Se appears to be more closely related to that of α-monoclinic Se than trigonal Se. However, the band structure of the crystalline forms of Se are probably not too different, since the nearest neighbor environments are essentially the same. Using the Se form factors and matrix elements calculated by Sandrock,[414] and assuming an appropriate two-particle correlation function, Kramer et al. evaluated the electronic band structure and lifetimes, and used these to estimate the imaginary part of the dielectric constant from 2 eV to 10 eV. In this range, agreement between the theoretical and experimental[415,416] ϵ_2 is quite good, particularly with regard to reproducing the two observed broad peaks near 4 eV and 8 eV, and the washing out of the crystalline peaks[417,418] near 3.2 eV and 6.5 eV. The agreement between theory and experiment is not very good near the 2.0 eV edge. This probably comes about because, as previously indicated, the structure near the edge is dominated be Se_8 ring transitions, and only chain-like short-range order was considered in the calculation. In addition, the assumption of equivalent densities of states in amorphous and trigonal Se precludes the possibility of the calculation reflecting any effects due to band tailing.

Kramer et al.[413] also calculated the joint density of states for transitions from the valence to the conduction band of trigonal Se and compared it with $\omega^2 \epsilon_2$, in order to obtain the energy dependence of the matrix elements for optical absorption. Instead of being nearly constant, a sharp maximum occurred near 4 eV. However, this maximum turns out to be due to umklapp enhancement,[419] and should not appear in the matrix elements appropriate to amorphous Se. For amorphous Se, relaxation of **k**-conservation requires that the joint density of states be replaced by a simple convolution, and the crystalline matrix elements be modified to suppress the effects of umklapp enhancement. Kramer et al.[413] showed that the assumption of energy-independent matrix elements leads to a greatly overestimated ϵ_2, particularly in the 4 eV region. These results cast some doubts on the procedures used in analyzing the photoemission data for amorphous Ge[247] discussed in subsection A.

5. Conclusions

Perhaps due to the more extensive experimental investigations, more appears to be understood at present about amorphous Se than about either Ge or Si. This is surprising in view of our relative insights into the corresponding crystalline materials. Furthermore, an *a priori* judgment would be that amorphous Se should be much more complex than Ge and Si because of the multiplicity of crystalline forms in the case of Se, each with a somewhat different short-range order. Perhaps the most important difference between the amorphous forms of Se and those of Ge and Si is that in the latter, little or no photoconductivity has been observed to date. On the other hand, photoconductivity in amorphous Se is readily detectable, primarily because of its high dark resistance, and it is because of this fact that we understand as much as we do about this complex material today. The following conclusions appear to be based on relatively firm ground at present.

1. Amorphous Se consists of both Se_8 rings and polymeric Se chains, although the percentages of each is still an open question. It is possible that the fractional amount of rings and chains is temperature dependent, and likely that they vary with the method of preparation of the material. But because of the fact that Se_8 rings are disjoint, amorphous Se is not a random-network glass, as amorphous Ge and Si appear to be. This arises primarily because of the fact that Se is a partially covalent, partially Van der Waals material, whereas Ge and Si are totally covalent. A purely covalently bound random network of Se atoms cannot fill three-dimensional space.

2. Intrinsic conduction characterizes liquid Se, the activation energy being about 1.1 eV. This behavior extrapolated to low temperatures appears to give the observed electrical conductivity of very pure amorphous Se. Intrinsic conduction can also be observed in compensated amorphous Se. The room-temperature conductivity is then approximately $10^{-17} \Omega^{-1}$ cm^{-1}.

3. The mobility of free holes in the valence band dominates that of free electrons in the conduction band. The free-hole mobility may be of the order of 20 cm^2/V-sec.

4. The electrical conductivity is extremely sensitive to oxygen impurities. Introduction of oxygen leads to relatively deep p-type acceptor levels.

5. A large density of localized states, of the order of 10^{20} cm^{-3} eV^{-1}, exists near the band edges in most samples. These appear not to be

band tails due to disorder, but rather acceptor and donor levels due to impurities or structural defects.

6. The hole conduction is predominantly band-like. The drift mobility is trap-limited, the traps most likely being the acceptor levels discussed above.

7. Hopping conduction can be observed in ac conductivity experiments. This conduction probably occurs within the acceptor levels rather than in any intrinsic localized states. The density of states at the Fermi energy is of the order of $10^{18}\ cm^{-3}\ eV^{-1}$.

8. High-field effects, such as space-charge-limited currents and Poole-Frenkel emission are readily observable in amorphous Se. This is still another indication of a significant density of traps in the material.

9. The optical absorption edge is about 0.6 eV below the photoconductivity edge. The non-photoconductive absorption does not represent the effects of band tailing, but rather of Frenkel-type excitonic transitions of Se_8[7] rings.

10. The density-of-states gap and the mobility gap of amorphous Se are both approximately 2.1 eV, indicating that extensive band tails due to positional disorder do not exist in the material. This is a significant conclusion, for it implies that the potential fluctuations in a particular amorphous material, even one with somewhat ambiguous short-range order, need not be very large. The explanation of this result most likely lies in the fact that both the Se-Se bond lengths and the bond angles in the monoclinic forms of Se and in trigonal Se are very similar. Thus, the short-range order in amorphous Se, monoclinic Se, and trigonal Se are essentially equivalent, a fact experimentally documented by the similarity of the optical absorption spectra of all of these materials.

V. COVALENT AMORPHOUS SEMICONDUCTORS

In this section, we shall be concerned with non-elemental amorphous solids in which the interatomic bonding is primarily covalent. These materials can be divided into two groups – stoichiometric binary compounds, such as As_2Se_3, Sb_2Te_3, or GeTe, which are relatively simple, and non-stoichiometric, multicomponent alloys, such as chalcogenide, arsenide, or boride glasses. Rather than be exhaustive, we shall deal with only a few examples, As_2Se_3, GeTe, and the multicomponent chalcogenide glasses.

A. As_2Se_3

1. Structure

As_2Se_3 crystallizes in a monoclinic layer structure,[420] in which each As atom is bonded to three Se atoms in a flat triangular-pyramid arrangement, three Se atoms forming the base and an As atom the apex. Each Se atom is part of two pyramid bases, thus fulfilling all of the covalent bonding requirements of both the As and Se atoms and extending the structure throughout a two-dimensional space. A projection of this layer onto a plane is shown in Figure 22a.[421] The layer structure can be looked at as puckered hexagonal 12-membered rings, linked together by Se atoms. The nearest-neighbor As-Se distance is 2.36Å. Since all the covalent bonding requirements are fulfilled within a single layer, the layers are held together by a relatively weak Van der Waals

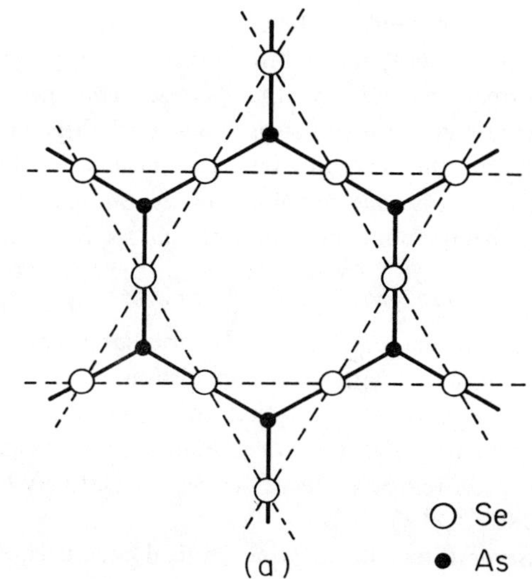

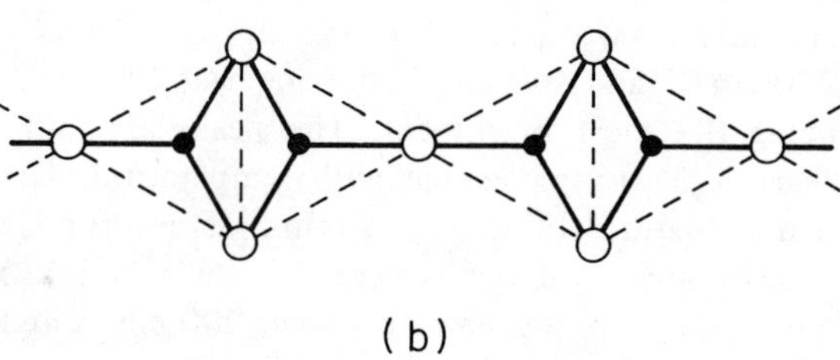

FIGURE 22. Two possible structures for As_2Se_3: (a) layers; (b) chains (Owen[421]).

attraction. The smallest interatomic distance between layers is consequently 3.66 Å, about 55% larger than the As-Se bond length within the layers.[422] Zallen et al.[422] point out that the interlayer distance is about 8% smaller than the sum of Pauling's[423] Van der Waals radii, indicating a weak component of covalent bonding exists between the layers. Nevertheless, the material is very easily cleaved parallel to the layer planes.[424] In addition to the layer structure, a one-dimensional chain-like structure is possible for As_2Se_3 by allowing two $AsSe_3$ pyramids to share a common edge, as shown in Figure 22b, rather than just a corner.[421] No allotropic crystal form has been found based on such a chain arrangement; however, it is a possible candidate for a component of the structure of amorphous As_2Se_3.

Vaipolin and Porai-Koshits[425-428] have carried out extensive x-ray diffraction studies of amorphous As_2Se_3 and related glasses, and have found evidence that the short-range order is essentially the same as in the crystal, the $AsSe_3$ pyramids being preserved with only small distortions. The long-range disorder is obtained not by cross-linking layers nor by formation of chain-like polymers but primarily by small deformations within the layers. The nearest-neighbor As-Se distance is 2.45 Å, slightly larger than in the crystal, but the nearest-neighbor interlayer distance remains 3.66 Å. Similar results have been found for amorphous As_2S_3, whose crystalline form is isomorphic with As_2Se_3.[429,430] Viscosity[431,432] and low-temperature specific-heat[433] measurements are also consistent with these conclusions.

Because of the fact that the interlayer forces are primarily Van der Waals interactions, the glass transition temperature of As_2Se_3 is relatively low, 460K.[432-434]

Several sets of infrared optical measurements have been performed on amorphous As_2Se_3.[435-440] The main absorption bands of crystalline As_2Se_3 are near 100 cm^{-1}, 220 cm^{-1}, 250 cm^{-1}, and 460 cm^{-1}, the one near 220 cm^{-1} being the most prominent. The peaks are fairly sharp and vary somewhat with temperature. The major features of the absorption of amorphous As_2Se_3 are broad bands near 100 cm^{-1} and 480 cm^{-1} and a strong band between 200 cm^{-1} and 275 cm^{-1}. These bands are relatively temperature insensitive, show no fine structure,[440] and in fact, appear to represent just a broadening of the crystalline bands due to disorder. The main band near 220 cm^{-1} can be associated with the As-Se stretching frequency and the band near 100 cm^{-1} with the bending frequency; the absorption near 460 cm^{-1} is most likely the second harmonic of the former. Thus, there is no doubt that the $AsSe_3$ pyramids are preserved in the amorphous material.

2. Electrical Properties

Electrical conductivity of both bulk samples and thin films of amorphous As_2Se_3 has been measured often.[441-448] Bulk samples exhibit a resistivity of the order of 10^{12} Ω-cm at 300K, and conductivity is ohmic for applied fields of up to 10^5 V/cm. In the temperature range 300 to 450 K, the conductivity obeys the ordinary semiconductor relation,

$$\sigma = \sigma_o e^{-E_A/kT} ,$$

with a temperature-independent E_A generally in the range 0.85 to 1.00 eV, although a value as low as 0.60 eV has been reported.[442] Thin films with gold electrodes show ohmic conductivity for fields of less than 2 x 10^4 V/cm. Room-temperature resistivity has been reported in the range 3 x 10^{10} - 3 x 10^{11} Ω-cm. Annealing in argon at 400K for about 20 hours causes the room-temperature resistivity to approach 1.5 x 10^{11} Ω-cm. This can represent either an increase[447] or a decrease[448] relative to the as-deposited resistivity, depending on the initial value. As opposed to the case for bulk samples and thick films, the activation energy for thin films is not independent of temperature in the range 100 to 400K. The value of E_A increases from 0.1 eV to 1.5 eV as the temperature increases, and no intrinsic range appears.[447] The data on the thin films even after annealing, do not obey a $T^{1/4}$ law, Equation 28, although such a fit is substantially better than the assumption of a temperature-independent E_A. Thin films definitely do not exhibit the same type of behavior as bulk samples, and this could indicate a different structure or even composition in the films. Aluminum electrodes are blocking on As_2Se_3 films, producing variable room-temperature resistivities of the order of 1-3 x 10^{13} Ω-cm; annealing can bring about ohmic regions. Crystalline As_2Se_3 shows a very anisotropic behavior, the conductivity parallel to the covalently bound layers being a factor of 10^2 greater than that perpendicular to the layers.[447] The room-temperature resistivity parallel to the layers is about 3 x 10^{10} Ω-cm, and the activation energy is 0.92 eV, independent of temperature. Thus, the crystalline-to-amorphous transition in As_2Se_3 removes the anisotropy and leaves the carrier mobility, but essentially preserves the energy gap.

In the liquid state, the activation energy is approximately 1.5 eV from 700K to 1200K, and is independent of temperature.[449,450] Below 700K, a break in the conductivity data exists, the activation energy falling to about 1.0 eV.[450] No further break occurs at the glass transition temperature,[451] and the conductivity behaviors in the glassy and liquid regions match up quite well, as might be expected.

It has been reported that the electrical conductivity of amorphous As_2Se_3 and similar materials is generally insensitive to small concentrations of impurities,[452] in accordance with theoretical predictions,[453,454] although the statement is accurate only in comparison with ordinary crystalline semiconductors. In fact, it has been found that Ge monotonically decreases the conductivity, with 10% Ge producing a conductivity drop of a factor of 30,[455] that small amounts of I (~1%) increases the conductivity by a factor of 10, but larger concentrations bring about a more gradual decrease,[455] that about 1% Cu increases the conductivity by a factor of 100,[456-458] but similar concentrations of Ag increase the conductivity only by a factor of 4,[458] that 0.2% Ga and In increase the conductivity by factors of 10 and 2, respectively, and that 0.2% Tl decreases the conductivity by a factor of 25; further amounts of Tl, however, increase the conductivity sharply, 37% Tl producing an increase of a factor in excess of 10^7 relative to pure As_2Se_3.[459-461] In each case, the changes in conductivity induced by the impurity appear to result from changes in the activation energy for conduction.[458,459] On the other hand, Haisty and Krebs[450] found that even large deviations from stoichiometry produced changes in conductivity of less than a factor of 10, provided the composition was within the glass-forming region (less than 60% As).

The thermoelectric power of amorphous As_2Se_3 is positive,[442,462] indicative of primarily p-type conduction. The temperature dependence obeys Equation 17 quite well, although the data then yield a separation of only 0.5 eV between the Fermi energy and the valence-band mobility edge and a negative value for a (a = -3). The former value is inconsistent with the activation energy for conduction; the latter is theoretically impossible. Owen[421,462] suggested that these paradoxes can be resolved by assuming both holes and electrons contribute to conduction, and concluded that the mobility ratio was $\mu_e/\mu_h \approx 0.27$. This is valid only if the Fermi energy is in the center of the mobility gap, an unlikely situation if there is significant band tailing. Edmond[443] assumed that the mobility gap decreases linearly with increasing temperature, and used the thermoelectric power data to deduce a value for the mobility gap under the further assumptions that a=0 and the Fermi energy is in the center of the gap. It is more reasonable to simply assume the separation between the valence-band mobility edge and the Fermi energy varies with temperature as

$$E_f = E_{fo} - \beta T \quad , \tag{39}$$

in which case the choice a = 4, the value appropriate when ionized-impurity scattering[463] predominates, yields $\beta \approx 6 \times 10^{-4}$ eV/K for the temperature coefficient of E_f. This is the same order of magnitude as the experimental temperature coefficient of the optical gap in amorphous As_2Se_3, 5.6 x 10^{-4} eV/K.[464] The resulting value for E_{fo} of 0.05 eV is consistent with the corresponding electrical activation energy of 0.6 eV if the mobility increases with increasing temperature in the 300 to 400K range investigated. This increase is also consistent with the predominance of ionized-impurity scattering.[465]

A similar analysis for the thermoelectric-power data in liquid As_2Se_3[443] results in a value for the separation of the Fermi energy and the valence-band mobility edge of 1.2 eV at T=0. If we assume that optical-phonon scattering dominates in this temperature range, then we can conclude that the temperature coefficient, β, is 13 x 10^{-4} eV/K, consistent with the corresponding coefficient of the optical gap. The observed activation energy for conduction in this temperature region is 1.1 eV, implying consistency with the thermoelectric-power data if the mobility decreases exponentially with increasing temperature. The latter is consistent with the predominance of optical-phonon scattering, although the optical-phonon energy calculated from the reststrahlen absorption is only 0.03 eV, rather than the 0.1 eV required.

However, this discrepancy results only from small differences between large numbers, and is not significant, particularly in view of the uncertainty in the value of the activation energy for conduction. Note that for liquid As_2Se_3, a two-carrier analysis based on a Fermi energy in the center of the mobility gap does not give resonable results.[124] No Hall effect data have been reported for amorphous As_2Se_3. On the other hand, drift-mobility experiments have been performed often.[345,466-470] The most significant result of these observations is that the transit time of photoexcited carriers is dependent on the applied field,[345,468,470] so that a drift mobility of holes at fields of 10^5 V/cm is of the order of 2 - 4 x 10^{-5} cm^2/V-sec and appears to be proportional to the applied field.[470] The hole drift mobility is thermally activated, with an activation energy of 0.5 eV at fields near 10^5 V/cm.[468] The drift mobility of electrons is much smaller than that of holes.[466] The low magnitude and high activation energy of the hole drift mobility strongly imply that the mobility is trap-limited. The field dependence then represents a non-ohmic effect, such as Poole-Frenkel emission. Thus, we do not have even an order of magnitude estimate for the actual conductivity mobility of holes in the valence band of amorphous As_2Se_3 as yet.

The ac conductivity of amorphous As_2Se_3 has been measured extensively.[326,447,462,471-474] Often an ω^2 dependence of the ac conductivity is observed,[447,462,471] but this has been shown to be an artifact due to the contacts rather than the sample itself.[326,474] Street et al.[475] have shown how contact resistance and capacitance of thin-film sandwich structures can lead to an ω^2 increase of the ac conductivity in an intermediate frequency range. In this region, the apparent resistivity must be proportional to the thickness of the film, a fact that provides a straightforward test of the results. With Au electrodes the ac conductivity of amorphous As_2Se_3 is much larger than when Al contacts are used,[326,474] a result which indicates that losses are introduced by Au electrodes. This is consistent with the result that Au diffuses easily into As_2Se_3, but Al does not.[476] With Al electrodes, the ac conductivity is equal to the dc conductivity through frequencies of 10^3 Hz. At higher frequencies, the ac conductivity increases approximately proportional to ω through 10^{10} Hz.[473,474] There is some evidence of a saturation near 10^{10} Hz, although far infrared measurements in the 10^{12} to 10^{14} Hz range indicate the possibility that the ω dependence may extend up to the region in which phonon absorption begins to contribute losses.[473] Some of these results are shown in Figure 23. Equation 29 suggests an increase of ac conductivity approximately proportional to ω, due to hopping in the vicinity of the Fermi energy. An upper bound for the density of states at the Fermi energy can be estimated from the minimum $\sigma(\omega)$ measured[474] by using Equation 29 and taking the spatial extent of the localized wave functions, a, equal to an interatomic separation, 3Å. The observed parameters then give $g(E_f) < 2$ x 10^{18} cm^{-3} eV^{-1}. This value is very sensitive to the value of a in Equation 29; however, an increase of a from 3Å to 5 Å reduces the upper limit for g (E_f) to the order of 10^{17} cm^{-3} eV^{-1}. Pollack[477] recently derived a rigorous expression for the ac conductivity for uncorrelated hopping in the vicinity of the Fermi energy. The result is similar to Equation 29 except for a numerical factor of $\pi^2/23$, which serves to increase the upper limit for $g(E_f)$ to approximately 5 x 10^{18} cm^{-3} eV^{-1}. It would be worthwhile to check that the ac conductivity is proportional to the temperature, as Equation 29 predicts, but no experimental information has yet been reported on the temperature dependence in the region where $\sigma(\omega)$ is proportional to ω.

It would appear that the ac conduction at frequencies larger than 10^3 Hz is predominantly due to thermally activated hopping in the vicinity of the Fermi energy. However, this does not imply anything about the predominant mechanism for dc condition. However, Owen[478] has pointed out that the proportionality of $\sigma(\omega)$ to ω could be viewed as the *high* frequency limit of a dielectric loss spectrum, in which case the temperature dependence can be used to estimate the activation energy of the relaxation time.[479] For amorphous As_2Se_3, this activation energy is essentially the same as that for dc conduction. In addition, Main and Owen[480] found that the current-noise spectrum of some chalcogenide glasses is of the 1/f type, which could result from a distribution of lifetimes for excitation and trapping of carriers. If so, just the same distribution of lifetimes is required as the distribution of relaxation times necessary to obtain the observed $\sigma(\omega)$. Owen concluded that it is possible that the dc and ac processes fundamentally represent the same con-

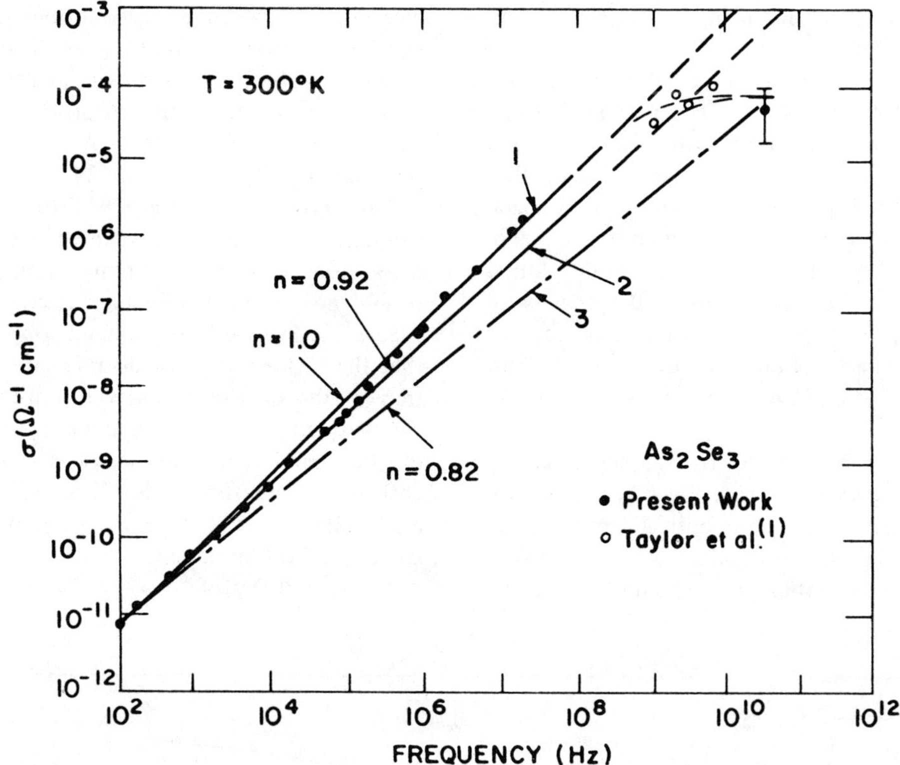

FIGURE 23. Frequency dependence of the electrical conductivity of amorphous As_2Se_3 at 300 K (Lakatos and Abkowitz[326]).

duction mechanism. This argument is somewhat weak, since there is no proof of a correlation between current noise and ac conductivity in these materials and several causes of 1/f noise other than carrier generation and recombination exist.[481]

Agarwal and Fritzsche[482] were not able to detect an EPR signal in amorphous As_2Se_3, despite the fact that the equipment was sensitive to 10^{14} cm^{-3}. Even at temperatures as low as 1.5 K, and with photo-excitation of free carriers, no signal was observed.[483] Thus these results indicate that the relaxation time is probably extremely short, rather than that no free spins exist in amorphous As_2Se_3. Tauc et al.[484] measured the magnetic susceptibility of the very similar material, amorphous As_2Se_3, and found a strong Curie term that is essentially absent in crystalline samples. The strength of this term corresponded to 6 x 10^{17} unpaired spins/cm^3. Thus, a relatively large free-spin density exists in amorphous As_2Se_3, and these arise primarily because of the long-range disorder. This could be understood structurally in terms of broken bonds or electronically in terms of overlapping band tails with a concomitant redistribution of the occupation of the valence and conduction bands. It is difficult to conclude the order of magnitude of the density of states at the Fermi energy from the total unpaired-spin density, since only the spins of the redistributed electrons are free, and these peak at the Fermi energy in the CFO model (see Figure 7). However, the magnetic susceptibility results are consistent with the upper limit of 5 x 10^{18} cm^{-3} eV^{-1} placed on $g(E_f)$ in amorphous As_2Se_3 by the ac conductivity measurements.

3. Optical Properties

Optical absorption experiments have been performed on amorphous As_2Se_3 for photon energies up to 14 eV.[443,447,485-487] Although the results obtained on different samples are similar, some discrepancies remain. The results near the edge are often interpreted in terms of Urbach's law,[488]

$$\alpha = \alpha_0 \, e^{\gamma(\hbar\omega - E_0)/kT} , \qquad (40)$$

a relation obeyed in many crystalline semiconductors.[489]

Urbach behavior produces a tail in the absorption edge, often with a characterstic 0.05 eV slope at room temperature, and is not fully understood at present.[490] However, it is quite clear that crystalline As_2Se_3 does not exhibit an Urbach edge,[486] and studies of the temperature dependence of the spectra of amorphous and liquid As_2Se_3 show a real red shift of the edge with increasing temperature, not just an increase of slope as suggested by Equation 40. [443,486] Thus, Urbach's law does not seem to be a factor in As_2Se_3.

Plots of $(\alpha \hbar \omega)^{1/2}$ vs. $\hbar\omega$, as suggested by Equation 34, produce straight lines in the vicinity of the edge, and result in an optical gap in the 1.7 to 1.8 eV range for amorphous As_2Se_3.[421,486] The optical gap is thus approximately twice the electrical activation energy, although, as emphasized in section III, there is no firm theoretical connection between the two quantities for disordered materials. The absorption spectra of crystalline and amorphous As_2Se_3 and related materials are shown in Figure 24.

The most striking feature of these data is the similarity of the corresponding crystalline and amorphous spectra. This point is emphasized by recent measurements of Zallen et al.[486] shown in Figure 25. For both As_2Se_3 and As_2S_3, it appears that the major effects of disorder are the elimination of the dichroism and van Hove singularities characteristic of the crystal and a slight broadening of the band tail. The energies at which the absorption coefficient is 2 x 10^4 cm^{-1} are essentially identical for corresponding crystalline and amorphous materials.

Grant and Yoffe[463] have measured the pres-

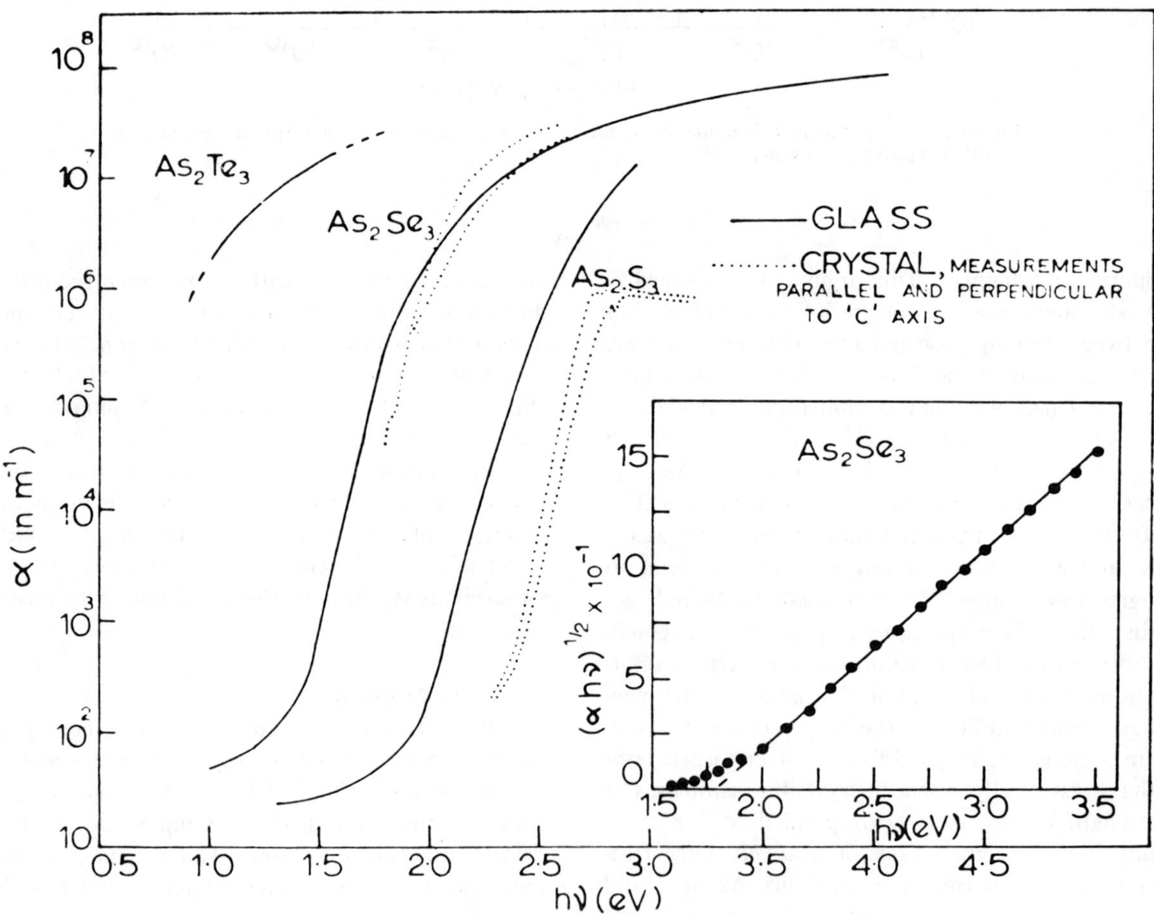

FIGURE 24. Optical absorption as a function of photon energy in amorphous As_2S_3, As_2Se_3, and As_2Te_3 in comparison with the corresponding crystalline forms (Owen[421]).

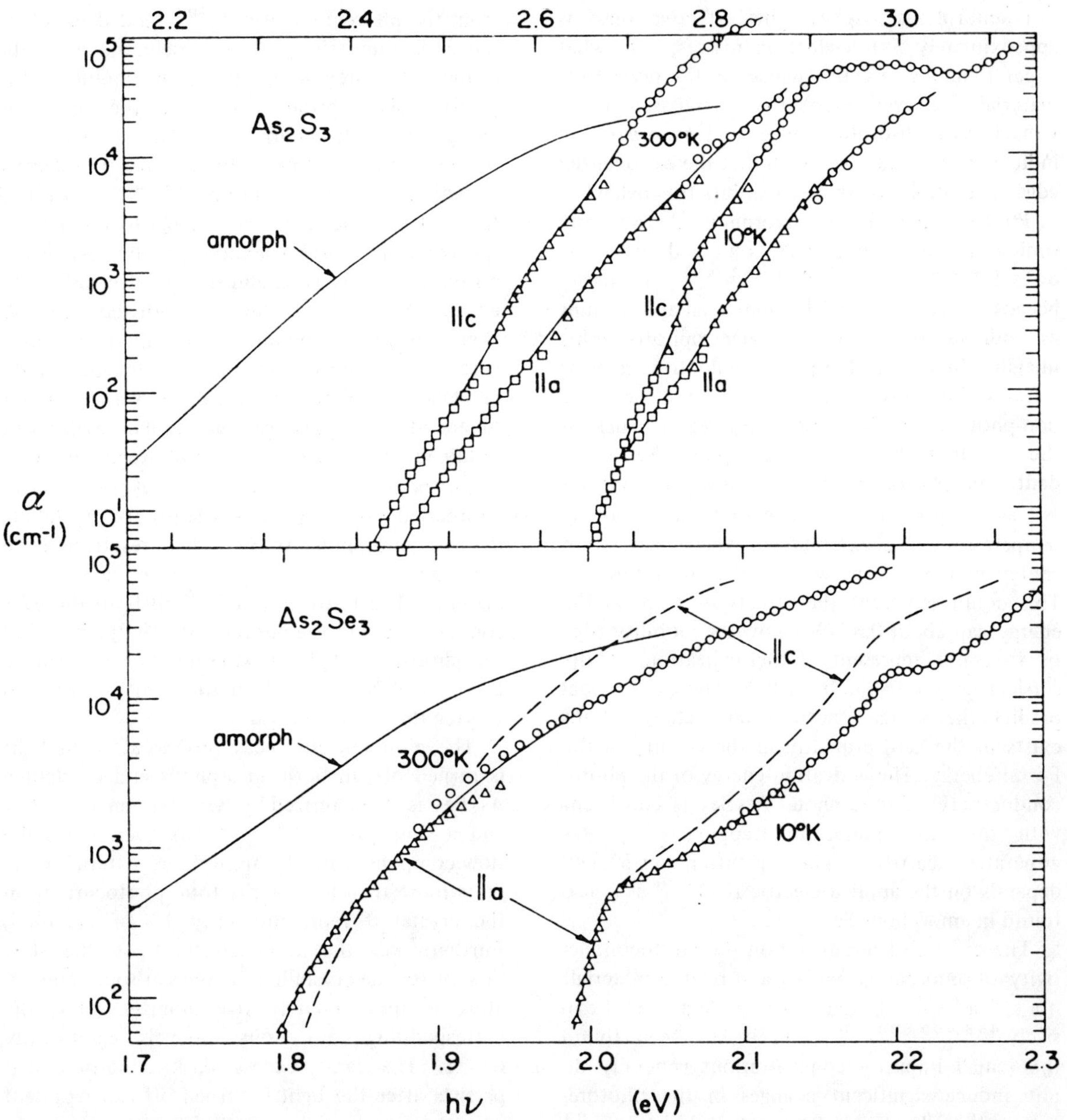

FIGURE 25. Optical absorption as a function of photon energy for corresponding crystalline and amorphous forms of As_2S_3 and As_2Se_3 (Zallen et al.[486]).

sure dependence of the optical absorption edge of crystalline and amorphous As_2Se_3, and found the edges decrease with increasing pressure. At 274K, the edge in the crystal decreases by 1.4 x 10^{-5} eV/bar, while the edge in the amorphous material decreases at about half this rate, 0.8 x 10^{-5} eV/bar. On the other hand, at 80K, the crystalline pressure coefficient is -1.0 x 10^{-5} eV/bar, while the corresponding amorphous value is -1.1 x 10^{-5} eV/bar. Thus, in amorphous As_2Se_3, the edge decreases with increasing pressure much more rapidly at low than at high temperatures.

Kolomiets et al.[491] studied electroabsorption in both crystalline and amorphous As_2Se_3, and observed a Franz-Keldysh shift of the absorption edge. The shift was proportional to the square of the applied electric field, but was not as large in the amorphous material as it was in the crystal. The theory of Franz[404] was then used to estimate the reduced effective mass of electrons and holes.

For amorphous As_2Se_3, this effective mass is approximately 3 free-electron masses, somewhat larger than the effective masses in the crystalline material. However, Franz theory is based on an Urbach-type absorption edge, and as previously indicated, it is not clear at all that the exponential edge in amorphous As_2Se_3 exhibits this behavior.

Photoconductivity in amorphous As_2Se_3 and similar materials has been observed in some detail,[438,441,447,469,485,492–498] primarily because of such potential commercial applications as vidicon tubes and in electrophotographic imaging. In general, the photoconductive response follows the absorption curve,[447] and no large non-photoconductive absorption region, such as the one that characterizes amorphous Se, is evident. The photoconductivity at low temperatures is an exponentially increasing function of temperature, and exhibits an average activation energy of 0.35 eV between 250K and 400K.[499] This could represent the effects of traps in the energy gap about 0.35 eV below the mobility edge or it could represent a thermalization of the carriers deep into the band tails. The CFO model predicts that a large concentration of charged traps exists in the gap, primarily in the vicinity of the Fermi energy. The activation energy of the photoconductivity of amorphous As_2Se_3 is consistent with this non-equilibrium trapping of photogenerated carriers. The quantum efficiency depends on the applied electric field,[470] as is also found in amorphous Se.

The effects of impurities on the photoconductivity of amorphous As_2Se_3 and related materials has also been studied extensively,[458,459,500–503] and it has been found that small impurity concentrations generally do not induce significant changes in the photoresponse.[500,501] Kolomiets and Rukhlyadev,[503] however, found that Ge impurity results in a blue shift of the photoconductivity edge, with 0.6% Ge inducing a 6% shift, and that Sn impurity leads to an initially small red shift. Recently, Kolomiets et al.[458,459] found that Ag, Cu, Ga, In, and Tl initially increase the energy of the photoconductivity edge in amorphous As_2Se_3, but, in each case, large concentrations lead to a maximum and eventual decrease in edge position. These photoconductivity edge shifts do not correlate with the corresponding shifts in absorption edge. Kolomiets et al. also noted that the product of mobility and carrier lifetime, which can easily be calculated from the photoconductivity,[504] tended to follow the dark conductivity rather closely, indicating the changes are predominantly in the mobility. Impurities also appear to change the activation energy of the photocurrent, an effect which can be looked at as a shifting of the average trap depths.

Kolomiets and Raspopova[495,496] measured the dark conductivity and the photoconductivity spectrum of amorphous and crystalline As_2Se_3 as a function of pressure, and found a decrease in the average activation energy for both dark conduction and photoconduction with increasing pressure. For the amorphous material but not for the crystals, the relative photocurrent was essentially independent of pressure, and similar results were obtained for the decay of the photoconductivity.

Furthermore, the magnitude of the relative photoconductivity was much larger for crystalline than for amorphous As_2Se_3. These results indicate the importance of trapping in the amorphous material. The traps appear to limit both the dark conductivity and the photoconductivity, as well as the photocurrent decay kinetics, as evidenced by the parallel behavior of all three with respect to varying the applied pressure.

The photocurrent decay process after the light is turned off, in both amorphous and crystalline As_2Se_3 is characterized by two components, a fast and a slow process.[496,499] However, while the slow component for the amorphous material represents more than half of the total photocurrent, in the crystal it represents only 1% of the total. Furthermore, the time constant of the slow process for the crystalline samples at room temperature is approximately five minutes, while the corresponding amorphous value is significantly longer. The large excess dark condition that persists after the light is turned off can represent only the effects of deep traps. These can be due to broken bonds resulting from structural defects or impurities, or they can arise intrinsically if a CFO-type band overlap exists.

In order to further study the nature of the traps, Kolomiets et al.[495,505] investigated thermally stimulated conductivity (TSC) in crystalline and amorphous As_2Se_3. The TSC curves for the amorphous material show a single peak in the vicinity of 300K. Kolomiets et al. calculated the trap energies by using several different theoretical models,[506–509] and obtained values centered in the 0.5 to 0.6 eV range. The occupied trap density was estimated as approximately 10^{17} cm^{-3}. Step-heating experiments[509] indicated a continuous distribution of traps from 0.35 eV to 0.75 eV from the mobility edge.[505] On the other hand, in crystalline As_2Se_3, Kolomiets et al. observed four distinct peaks in the TSC curves, indicating discrete trapping levels at given energies from 0.25 eV to 0.65 eV from the band edge.

Kolomiets et al.[510] have recently observed photoluminscence in crystalline and amorphous As_2Se_3 and found that both exhibited similarly shaped single peaks at 77K, with a 0.4 eV half width. However, the peak in the crystal occurred at 1.1 eV, while the peak in the amorphous material was centered at 0.9 eV. In each case, the photoluminescence occurred almost exclusively with photon energies less than the optical gap. At 120K, a secondary peak in photoluminescence was observed in the vicinity of the optical gap, showing direct band-to-band recombination does take place. Similar results had previously been found in $As_2Se_3 \cdot As_2Te_3$.[511] There appears to be no question that the photoluminescence is dominated by transitions to deep traps in the gap in both crystalline and amorphous material. It is possible that the crystalline traps are due to structural defects and impurities, while the traps in amorphous As_2Se_3 arise from a CFO-type band-tail overlap. However, Kolomiets et al.[510] determined that an increase in Se concentration, which produces nonstoichiometric compositions, leads to an increase in photoluminescence intensity. This is most simply explained by an increase in traps due to a greater concentration of dangling bonds, and is thus perhaps an indication of the existence of ordinary trapping levels in amorphous As_2Se_3.

Fischer et al.[512] observed photoluminescence in amorphous $As_2Se_3 \cdot 2As_2Te_3$ films from 2K to 150K. In this material, the mobility gap is about 1.0 eV and the optical gap is 1.1 eV.[513] A single photoluminescence peak occurred near 0.6 eV, and its shape was essentially independent of temperature.[512] The luminescence intensity was thermally activated below 100 K, with an activation energy of 0.03 eV; above 100K, the activation energy appeared to increase to 0.10 eV. Weiser et al.[513] previously had concluded from the 0.2 eV activation energy of the photoconductivity at light intensities sufficiently high that bimolecular recombination processes predominate, that a "recombination edge" exists, at which the probabilities of thermalization and recombination are the same. The concept of a recombination edge is based on the observation that as thermalization proceeds below the mobility edge, the states become increasingly more localized and more spatially separated. Thus, the probability of connecting any two such widely separated states by emission of a single phonon becomes very small. The photoconductivity data then indicate that for carriers in states greater than about 0.2 eV beyond the mobility edge in amorphous $As_2Se_3 \cdot 2As_2Te_3$, the recombination rate exceeds the rate of phonon emission. Because of an exponential decrease in lifetime with increasing temperature in the low light-intensity region, a negative activation energy of -0.13 eV is observed in the photoconductivity. These values self-consistently yield 1.0 eV for the mobility edge, in agreement with the direct experimental results. Fischer et al.[512] used these values and a simplified model to properly predict the 0.6 eV peak in photoluminescence, although the dependence of peak position on light intensity was greatly overestimated. This could either indicate peaks in the density of states 0.2 eV from the respective valence and conduction band mobility edges, or be explained by the presence of ordinary traps separated by about 0.6 eV. If the concept is applicable, the value for the recombination edge in amorphous As_2Se_3 can be estimated from the activation energy of the photoconductivity[499] to be about 0.35 eV above the valence-band mobility edge. This figure is consistent with the photoluminescence peak appearing at 0.9 eV.[510]

Kolomiets[514] has summarized the energy spectrum of localized states in the gap of both crystalline and amorphous As_2Se_3, as determined by analysis of thermally stimulated conductivity (TSC), space-charge-limited currents (SCLC), induced photoconductivity (IPC), and optical quenching of photoconductivity (OQPC) experiments. The results are shown in Figure 26. The photoluminescence levels approximately 0.2 eV from the two mobility edges in the amorphous material could well be added to the sketch. The major differences between crystalline and amorphous As_2Se_3 according to Figure 26 are the greater trap density and more spread-out nature of the trap levels in the latter.

Kruglov and Zimkina[515] observed the ultrasoft x-ray emission spectrum of amorphous As_2Se_3 in the 30 to 60 eV range. They were able to identify the M_{IV} and M_V emission bands of both arsenic and selenium, and concluded that the valence band

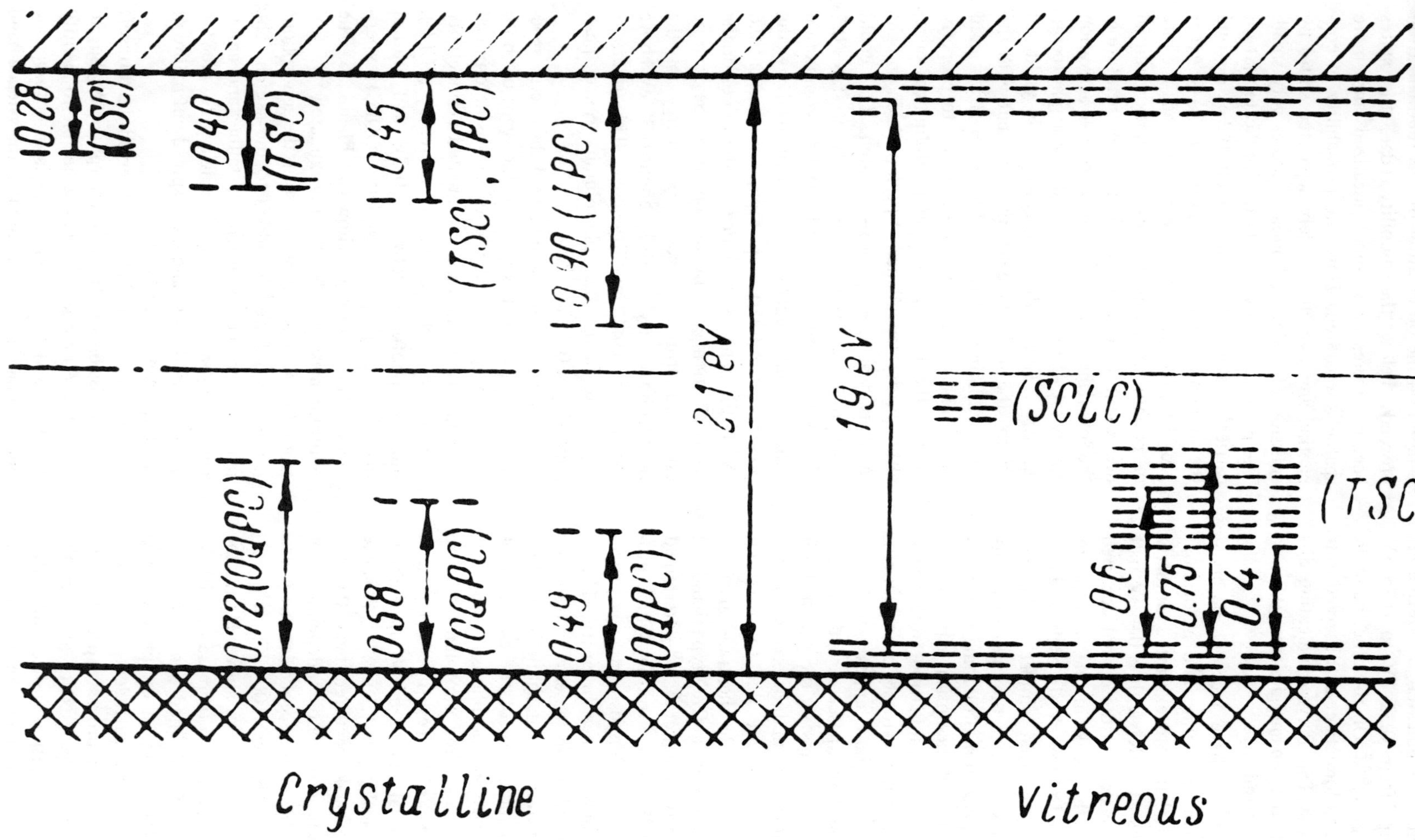

FIGURE 26. Energy spectrum of localized states in the gap for crystalline and amorphous As_2Se_3, as obtained from thermally stimulated conductivity (TSC), space-charge-limited conductivity (SCLC), induced photoconductivity (IPC), and optical quenching of photoconductivity (OQPC) experiments (Kolomiets[514]).

of the solid was approximately 7 eV wide. Thus, the outer bands in amorphous As_2Se_3 are of ordinary width, and the Anderson transition should not be a factor. Both the valence and conduction bands appear to be primarily of p symmetry, not surprising in view of the fact that the valence electrons of both As and Se are 4p electrons and the material is covalently bound. The band gap appears to about 1.1 eV, considerably lower than the optical gap of 1.8 eV. The material being studied was under constant bombardment by an electron beam and thus was at an effective temperature well above 300K. The gap does decrease with increasing temperature, but the observed coefficient[464] is such that it would take a 1000K increase over ambient to remove the discrepancy. If this were the case, the material would have liquefied; thus it is more likely that the result for the gap is somewhat inaccurate.

In general, the properties of materials very closely related to As_2Se_3, such as As_2S_3, As_2Te_3, Sb_2Se_3, and Sb_2S_3 are analogous to those of As_2Se_3, except for quantitative variations of the optical and mobility gaps.[516-527] An interesting result, however, is that in Sb_2Se_3, the short-range order of the amorphous material appears to differ significantly from that of the crystal.[528,529]

4. Conclusions

The large number of experimental papers on amorphous As_2Se_3 enable us to reach the following tentative conclusions.

1. The short-range order of the crystalline and amorphous solids is essentially identical, and consists of puckered 12-member rings as shown in Figure 22a. This structure satisfies all of the covalent bonding requirements of the constituent atoms, and still depends on Van der Waals interactions to fill three-dimensional space. However, the anisotropy of the crystal is removed in the amorphous phase.

2. The electrical conductivity of bulk glassy As_2Se_3 is characterized by a single activation energy of 0.9 eV over a large temperature range. No $T^{1/4}$ region has been observed as yet. The conductivity in thin films, on the other hand, depends critically on both the method of deposition and the electrodes. Annealing the films tends to even out the dispersion in values of the conductivity. Au generally provides ohmic contacts, while Al electrodes tend to be blocking. This effect, although arising from the fact that Au diffuses easily into As_2Se_3 even at room temperature, is an indication that the Fermi energy is not completely pinned in amorphous As_2Se_3.

3. Formation of non-stoichiometric glasses, with either excess As or excess Se, does not have much effect on the electrical conductivity. However, even small concentrations of certain impurities induce large changes in conductivity, and these changes are correlated with variations in the activation energy rather than arising from mobility effects. The decrease in activation energy could represent impurity lowering of the mobility gap, but could also be due to bond breaking and the consequent formation of deep traps in the gap. If the latter is the case, the traps must be distributed throughout a region of energy, rather than at a well-defined level, since the activation energy varies with impurity concentration.

4. Impurities also affect the photoconductivity in amorphous As_2Se_3, primarily by changing the drift mobility of photogenerated carriers. This is further evidence for the introduction of deep traps and a trap-controlled mobility. Traps appear to limit the dark conduction, the photoconduction, and the decay kinetics.

5. Hole conduction appears to dominate electron conduction, according to the results of thermoelectric-power and drift-mobility measurements. The drift mobility depends on the applied field, and is trap-controlled. An electronic effect, such as Poole-Frenkel emission, leads to the field dependence. Because of this, and the failure to detect a Hall voltage, no real estimate of the conductivity mobility has been extracted from the transport data. However, in amorphous As_2Se_3, a similar material, a lower limit of 0.3 cm^2/V-sec has been deduced from an analysis of photoconductivity.[524] The x-ray estimate of a 7 eV valence bandwidth indicates a band-like mobility should exist inside the mobility edges.

6. The kinetics of photoconductive decay are very different in crystalline and amorphous As_2Se_3, indicating the greater importance of charged traps in the latter material. Similar differences between crystal and glass have been observed in pressure experiments.

7. The optical gap is about 1.8 eV, essentially twice the activation energy for electrical conduction. This indicates that the Fermi energy is near the center of the mobility gap. The optical

absorption edge appears to be just a spread-out version of the crystalline edge.

8. The ac conductivity provides evidence of the existence of hopping conduction in the vicinity of the Fermi energy, although this appears to be dominated in the dc conductivity by free-hole conduction below the valence-band mobility edge at ordinary temperatures. The ac conductivity also provides an upper limit of about 10^{18} cm^{-3} eV^{-1} for the density of states at the Fermi energy. This upper limit is consistent with the lower limit for traps in the gap of 6×10^{17} cm^{-3}, deduced from magnetic-susceptibility experiments. No EPR signal has yet been observed, presumably because of a very short relaxation time.

9. There is a great deal of evidence for the existence of continuous distribution of traps in the energy gap. These can be observed in photoconductivity, thermally stimulated conductivity and photoluminescence experiments. It is not yet clear whether these traps represent primarily the effect of dangling bonds due to structural defects and impurities or primarily the result of the redistribution of electronic occupations necessary to bring about thermal equilibrium if a CFO-type band overlap exists. However, the weight of present evidence seems to favor the former explanation. The lack of pinning of the Fermi energy, the relatively small upper limit for $g(E_f)$, the lack of sensitivity of the photoluminescence peak to light intensity, and the values obtained for the trapping energies all point to amorphous As_2Se_3 as an ordinary highly compensated semiconductor.

B. GeTe

Although it has not been investigated with anywhere near the thoroughness of either As_2Se_3 or the multicomponent chalcogenide glasses, amorphous GeTe is worthy of a brief discussion, since it represents a material in which it has been established that the short-range order is *not* the same as that of the corresponding crystal.

1. Structure

GeTe crystallizes in a rhombohedral structure,[530-532] which is just a slight distortion of the NaCl structure. There is an elongation along one of the cube diagonals, resulting in an 88.4° rhombic angle, as opposed to the 90° angle in NaCl. The distortion has been ascribed to the effects of non-stoichiometry,[533] high carrier concentration,[534] the difference in the ionization potentials of Ge and Te,[535] and the fact that the atomic radius of Te is over twice that of Ge,[536] but the actual cause is still an open question.[537] In any event, GeTe crystallizes in a structure which is ordinarily associated with primarily ionic rather than primarily covalent bonding.[538] There is reason to believe that an effective charge of about +0.5e is present on the Ge atoms, with a corresponding negative charge on the Te atoms.[539] Above about 650K, GeTe undergoes an allotropic phase transformation to an ordinary NaCl structure.[540] The nearest-neighbor Ge-Te separation in either crystalline form is approximately 3.0A° and the coordination number for both Ge and Te is 6.

Amorphous GeTe cannot be obtained by quenching from the melt.[541] However, amorphous films can be prepared by vapor deposition onto a substrate whose temperature is held above 380K.[542,543] The density of these films is approximately 9% less than the crystalline value.[544] Although early work on Te-rich GeTe films[545] and on more stoichiometric films[543] were interpreted as indicating the basic equivalence of the short-range order of crystalline and amorphous materials,[546] recent radial distribution studies of Bienenstock et al.,[547] Betts et al.,[548] and Dove et al.[549] now clearly indicate that both the nearest-neighbor separation and the coordination number of amorphous GeTe differ from those of crystalline GeTe. The nearest-neighbor distance decreases from 3.0Å in the crystal to 2.7Å in the amorphous films. There is not yet complete agreement on the coordination number of amorphous GeTe. Dove et al.[549] concluded that the coordination number is essentially 4 and that the material was locally tetrahedrally ordered. Luo[550] obtained a coordination number of about 2 on Te-rich films, while Hilton et al.[551] arrived at a similar conclusion from studies on a ternary glass, $Ge_{15}As_{45}Te_{40}$. Betts et al.[548] found that the best fit to their data on several amorphous Ge-Te alloys occurred when the coordination numbers of the Ge and Te atoms differed. They proposed a "random covalent" model, in which each Ge atom is fourfold coordinated and each Te atom is twofold coordinated, so that all valence requirements of both atoms are locally satisfied. An example of this type of structure is shown in Figure 27. Some evidence for this model is that the nearest-neighbor separation, 2.7Å, is quite near

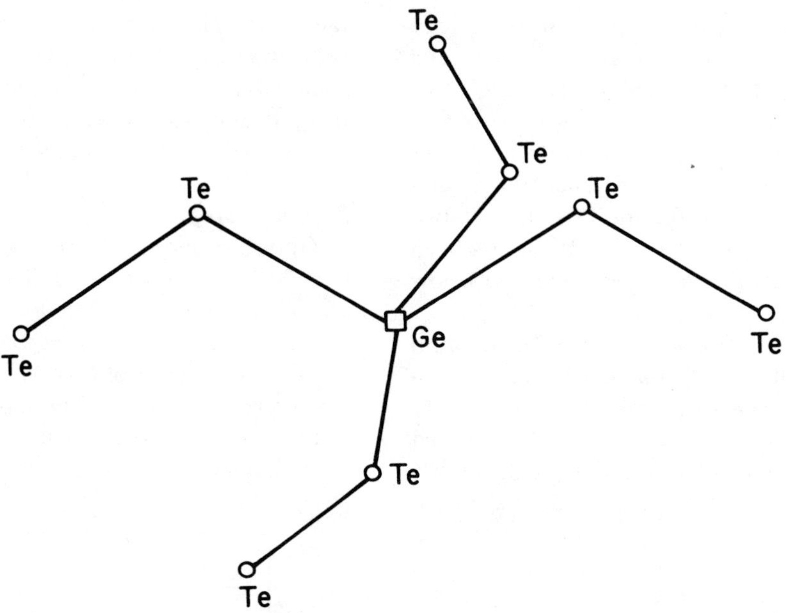

FIGURE 27. Sketch of the atomic arrangements in the vicinity of a Ge atom in amorphous Ge-Te alloys (Adler et al.[539]).

the sum of the Ge and Te covalent radii, 2.54Å.[552] The sum of the ionic radii of Ge and Te is 3.14Å, much nearer to the crystalline nearest-neighbor separation of 3.0Å, but far from the corresponding amorphous value. We can conclude that the amorphous material is much less ionic than the crystal. An analysis[539] of nuclear magnetic resonance studies indicates that the effective charge on a Te atom which has one nearest-neighbor Ge atom in amorphous Ge-Te alloys is -0.18e. Although this is roughly equivalent to the effective charge per bond in the crystal, the smaller coordination number in the amorphous material leads to the latter being much more covalent. Betts et al.[553] performed X-ray absorption studies of amorphous Ge-Te alloys and found that the Ge K absorption edge was similar to that of crystalline Ge, while crystallized material showed significant shifts relative to crystalline Ge. Betts et al. further found from x-ray photoemission experiments that the amorphous alloys exhibit only small changes in Ge core binding energies relative to crystalline Ge. On the other hand, crystallized material produced much larger changes, indicative of greater ionicity. Bienenstock et al.[554] studied the eutectic alloy, $Ge_{17}Te_{83}$, by both x-ray and neutron diffraction and found the radial distribution functions obtained were essentially identical, and consistent with the random covalent model. The evidence, at present, seems to point very strongly in favor of this model.

2. Electrical Properties

Crystalline GeTe is always a degenerate p-type semiconductor, with carrier concentration in the range 10^{20} to 10^{21} cm^{-3}, essentially independent of temperature.[555-559] The large carrier concentration is generally attributed to the presence of approximately 2% Ge vacancies, and a calculation of the energy necessary to remove a Ge atom from the crystal has indicated a vacancy concentration of 10^{22} cm^{-3} at equilibrium.[560] However, such a calculation is based on small differences between approximate large numbers and is generally quite inaccurate. Furthermore, a pure vacancy model cannot account for the dependence of carrier concentration on stoichiometry and heat treatment.[561] Antistructure, in the form of Ge atoms on Te sites, has been proposed as an alternative type of defect,[561,562] although the significant degree of ionicity of the Ge-Te bond in the crystal would appear to suppress large concentrations of isolated antistructure. Interstitials would seem to be a more reasonable minority defect for GeTe. Tunneling studies indicate that the energy gap of crystalline GeTe is quite small, of the order of 0.1

eV.[563,564] However, the optical absorption edge occurs in the 0.7 to 1.0 eV range,[565] because of a strong Burstein shift[566] due to the large free carrier concentration together with the **k** - conservation rule for optical transitions in the crystal. There is both experimental[567] and theoretical[568] evidence for the existence of two valence bands in crystalline GeTe. It has also been shown that GeTe becomes superconducting below about 0.4K.[569]

The resistivity of amorphous GeTe at room temperature is about a factor of 10^7 greater than that of crystalline GeTe.[562,570] The conductivity of the amorphous material is thermally activated from 77K to the crystallization temperature, 410K, with an activation energy in the 0.3 to 0.4 eV range, essentially independent of temperature. A typical result is shown in Figure 28. Conductivity is ohmic up to fields of about 2 x 10^3 V/cm at room temperature.[562] A p-type Hall effect has been observed, indicating a carrier concentration of the order of 10^{18} cm^{-3} and a Hall mobility less than 10^{-2} cm^2/V-sec.[562] It has been reported that the thermoelectric power is also p-type and about 900μV/K in magnitude, independent of temperature.[562] The lack of thermal activation energy in these measurements is some evidence that hopping conduction may predominate in amorphous GeTe,[571] but, as previously noted, thermoelectric-power measurements should be treated with caution.

The ac conductivity of amorphous GeTe is essentially independent of frequency at high temperatures (greater than 300K) and independent of temperature at high frequencies (greater than 10^5 Hz).[562] At low temperatures, the ac conductivity obeys an ω^n relationship, with n between 0.5 and 1.0. These results can easily be explained as the net result of two parallel conduction processes, a band-like mechanism with activation energy near 0.38 eV which predominates at low frequencies and high temperatures, and a temperature-insensitive hopping mechanism which predominates at high frequencies and low temperatures. There is evidence for saturation of the hopping conduction above 10^6 Hz, which indicates a very long hopping time, of the order of 10^{-6} sec.

Tunneling data have also been obtained,[562] and a broad minimum in dI/dV exists in the range -0.3V to +0.3V. This could indicate a mobility gap near 0.6 eV with the Fermi energy essentially in the center. However, in the absence of data on the temperature dependence of the tunneling conductance it is presumptuous to conclude much from the room-temperature results, which may not even represent real tunneling.

3. Optical Properties

Optical absorption experiments on amorphous GeTe[570,572] show an absorption edge which obeys Equation 34, and results in an optical gap of about 0.8 eV at 77K, decreasing to 0.7 eV at 300K. Typical plots of $(\alpha\hbar\omega)^{1/2}$ as a function of $\hbar\omega$ are shown in Figure 29. The absorption coefficient is about a factor of 10 smaller in amorphous than in crystalline GeTe throughout the entire range of

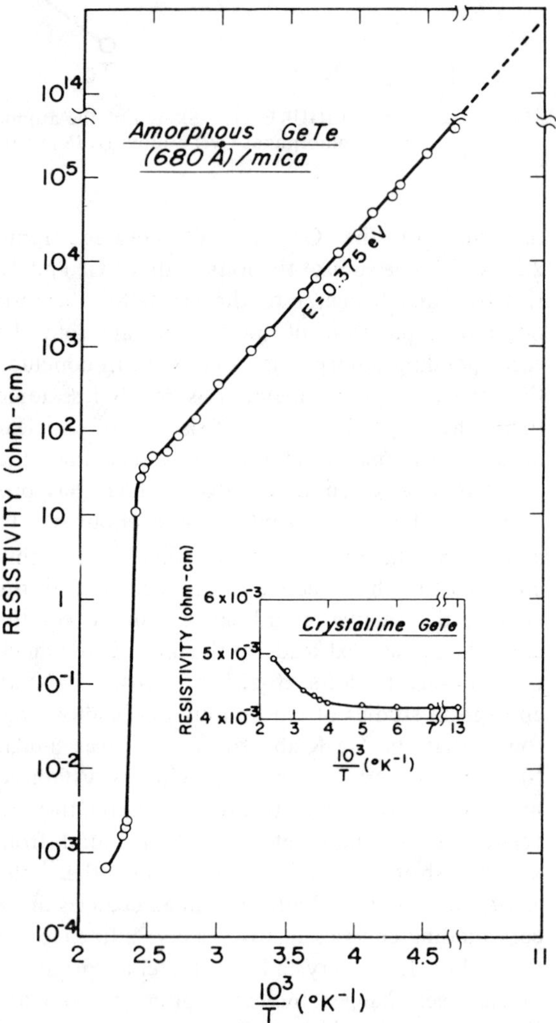

FIGURE 28. Resistivity as a function of temperature for an amorphous GeTe film of thickness 630A evaporated on mica at 300 K (Bahl and Chopra[562]).

photon energies shown. The actual position of the edge is nearly identical in crystalline and amorphous GeTe; however, since the edge in the crystal is strongly Burstein shifted and the short-range order is very different in the two phases, this is entirely accidental. The high-frequency dielectric constant is 11 in the amorphous material,[570] but 37 in the crystal,[565] underscoring the basic difference between the phases. The absorption edge in amorphous GeTe is exponential with a 0.07 eV tail, a typical value for both crystalline and amorphous semiconductors. However, as is clear from Figure 29, this behavior is not that of Urbach's law, since the slopes of the tails at 77K and 300K are identical. Howard and Tsu[570] assumed that the exponential tail of the optical absorption represented the effects of band tails in the density of states. The further assumptions of symmetrical valence and conduction bands, constant matrix elements, and parabolic bands above the mobility edges then yield a CFO-type band overlap with $g(E_F) \sim 10^{18}$ cm^{-3} eV^{-1}. The corresponding density of localized states in the mobility gap is 2×10^{19} cm^{-3}, of which approximately 10^{17} cm^{-3} should contribute charged traps at T = O. The latter figure indicates as average separation of about 200Å between the charged traps in the gap.

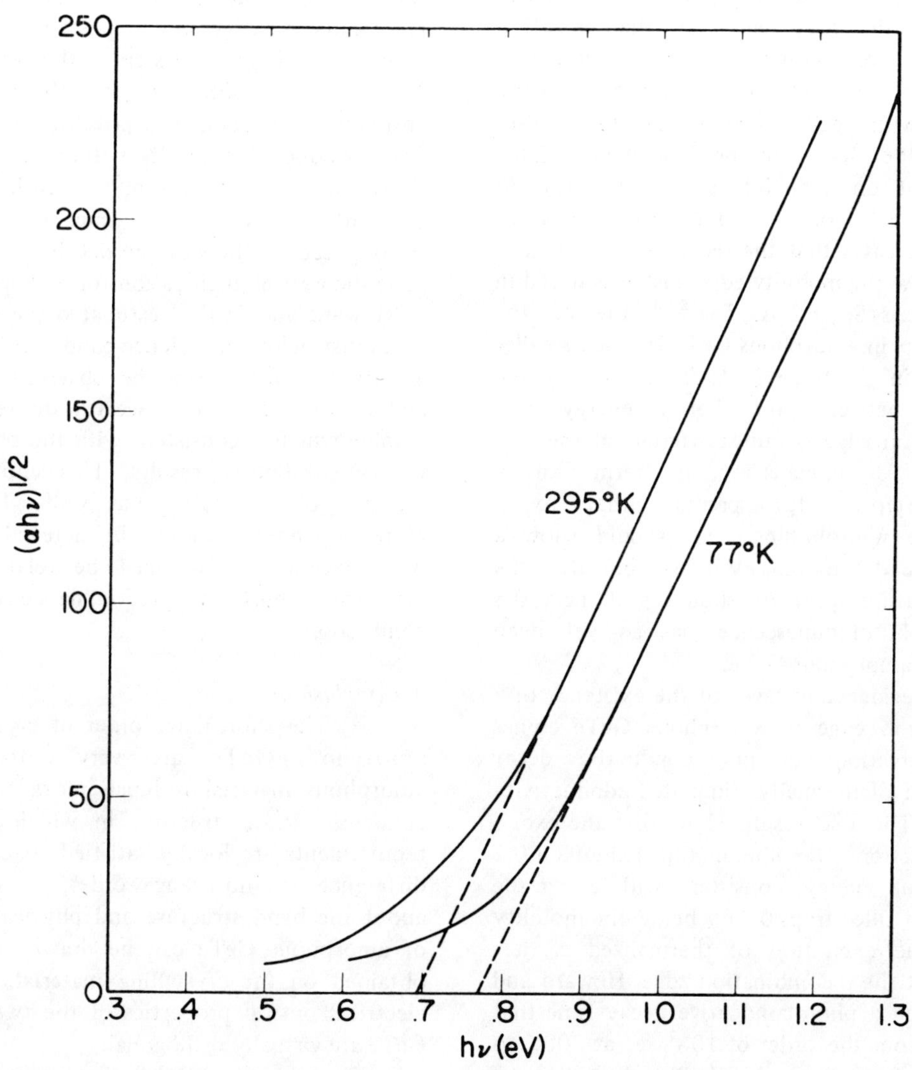

FIGURE 29. Square root of the product of optical absorption coefficient and photon energy for an amorphous GeTe film at two different temperatures (Howard and Tsu[570]).

As opposed to the situation in the crystal, photoconductivity is readily observed in amorphous GeTe films.[562,570] At low temperatures, the photoconductivity edge is near 0.6 eV, but low-energy tails begin to appear above 200K, and measurable photocurrent persists below 0.2 eV at room temperature.[570] The photoconductivity is proportional to light intensity at room temperature, but approaches a square-root behavior below 140K. This implies that the density of trapped carriers at room temperature is small compared to the thermally excited density, but the reverse is true at low temperature. Howard and Tsu[570,573] observed that all the photoconductivity results could be explained easily by postulating a recombination edge in the band tail, as discussed in sub-section A, below which states are sufficiently localized that thermalization proceeds only very slowly. If such an edge exists, the ratio of trapped to free carriers sharply increases with decreasing temperature, thus explaining the change of the dependence of the photocurrent on light intensity from linear to square-root behavior. A quantitative analysis of the data indicates that the recombination edge is 0.2 eV below the mobility edge, just as is found in amorphous $As_2Se_3 \cdot 2\ As_2Te_3$.[513] However, the mobility gap in amorphous GeTe is much smaller than that in $As_2Se_3 \cdot 2\ As_2Te_3$, so that the separation between the Fermi energy and recombination edge is quite different in the two materials. An explanation in terms of a recombination edge predicts that room-temperature photoluminescence should show a peak in the 0.3 to 0.5 eV range, but that the low-temperature spectrum should peak near 0.8 eV. No photoluminescence has as yet been reported in amorphous GeTe.

Further evidence in favor of the existence of a recombination edge in amorphous GeTe comes from observations of photoconductive decay kinetics and of thermally stimulated conductivity (TSC).[573] The TSC results show that the excess dark conduction after illumination exhibits a 0.2 eV activation energy, consistent with either the emptying of filled traps 0.2 eV below the mobility edge or the excitation of thermalized carriers frozen in at the recombination edge. Howard and Tsu observed a photoconductive decay time that increased from the order of 10^{-2} sec at 300K to the order of 10^5 sec at 100K. By subtracting off the long-time component of excess current at low temperatures, they obtained a second component with decay time of the order of 10^3 sec. The assumption of recombination edges in a CFO-type model implies the existence of two sets of trapping levels for each type of carrier, those at the recombination edge and those near the Fermi energy. Howard and Tsu associated the long-time component of the photoconductive decay with the emptying of the traps near E_F, and the shorter component with the equilibration of carriers trapped near the recombination edge. The long-time process should then exhibit a decay time constant which has the same activation energy as the conductivity, consistent with a change from 10^{-2} sec at 300K to 10^5 sec at 100K. On the other hand, the decay time constant associated with the equilibration of carriers at the recombination edge should possess only a 0.2 eV activation energy, consistent with a value of only 10^3 sec at 100K. Despite these impressive correlations of data, it is possible to explain the photoconductivity results without accepting the CFO model, by simply invoking several distributions of ordinary trapping levels in the gap. As opposed to the CFO model, however, there is little theoretical justification for such an approach.

Howard and Tsu[570] estimated the mobility of holes just below the valence-band mobility edge of amorphous GeTe from the observed $\mu\tau$ product and a choice of $\tau \sim 10^{-10}$ sec at room temperature, a value which is consistent with the photocurrent vs. light intensity results. This choice gives a mobility of 1 cm^2/V-sec at 300K. The temperature dependence cannot be determined in this way, except that it must be relatively weak, evidence against the predominance of hopping conduction.

4. Conclusions

1. The short-range order of crystalline and amorphous GeTe are very different. The amorphous material is much more covalent and condenses in a structure in which all valence requirements are locally satisfied. Because of this difference in short-range order, *no* conclusions about the band structure and physical properties of amorphous GeTe can be drawn from results obtained on the crystalline material. Indeed the electrical optical properties of the two phases of GeTe are virtually orthogonal.

2. Amorphous GeTe is a p-type semiconductor, and exhibits a thermally activated

conductivity with an activation energy of approximately 0.35 eV. The hole mobility is of the order of 1 cm^2/V-sec and is not a strong function of temperature.

3. The optical gap is near 0.7 eV, approximately twice the activation energy for electrical conduction. There is definite evidence for the existence of band tails from the optical spectrum, and a density of states at the Fermi energy of 10^{18} cm^{-3} eV^{-1} has been estimated. However, this result should be treated with some degree of caution, since a large number of crystalline semiconductors exhibit similarly shaped, but basically unexplained absorption edges.

4. There is definite evidence for two separate types of traps, one set distributed near the Fermi energy, the other about 0.2 eV below the mobility edge. Whether these are traps due to impurities and structural defects or arise naturally from extensive band tails comprised of states of ever increasing localization is not at all clear at the present time.

C. Multicomponent Chalcogenide Glasses

It is somewhat foolhardy to consider together the uncountable infinity of materials that make up the category of chalcogenide glasses, but the fact remains that certain unique modes of behavior appear to be generic to glassy systems containing three or more different chemical species including at least one of the chalcogen elements, S, Se, and Te. Chalcogenide glasses have been studied for over one hundred years,[574] but intensive investigation began only after the discovery that many of these materials transmit light well in the infrared, out to 12μ.[575-577] Research on the multicomponent chalcogenide glasses was spurred by the discovery that $Tl_2Se \cdot As_2Se_3$ possessed ordinary semiconducting electrical properties, with a room-temperature conductivity of the order of 10^{-9} Ω^{-1} cm^{-1} [578,579] Shortly thereafter it was found that amorphous $Tl_2Se \cdot As_2Te_3$ had a room-temperature conductivity of 10^{-3} Ω^{-1} cm^{-1}, larger than many crystalline semiconductors.[580] This initiated an intensive study of the regions of glass formation of ternary and quarternary chalcogenide alloys and their physical properties.[581-602] Halogen-containing chalcogenide glasses also were investigated[603-605] after it was discovered that they have desirable features for use as high-refractive-index immersion media.[606,607] Research on chalcogenide glasses continued at about the same level through 1968, when new prospects of commercial applications[1] spurred still greater efforts. The chalcogenide glasses today rank among the most investigated classes of semiconducting solids.

1. Structure

It is obviously an extremely difficult proposition to study the structure of ternary and quarternary glasses by means of x-ray or electron diffraction experiments, except to verify their amorphous nature by noting the absence of Bragg scattering peaks. Because of the simultaneous presence of many different covalent-bond separations in multicomponent alloys, the radial distribution functions are essentially uninterpretable. Consequently, few diffraction results have been reported in any but some of the simpler pseudobinary systems. In most of the binary alloys, with GeTe the most glaring exception, the short-range order of the corresponding amorphous and crystalline materials are the same. However, in the pseudobinary As_2Se_3-As_2Te_3 system, Vaipolin and Porai-Koshits[427] noted some differences in short-range order, and interpreted these in terms of relative covalency and ionicity of the constituent bonds. Studies of the PbS-As_2S_3 system[608] led to the conclusion that As_2S_3-rich glasses are similar in structure to amorphous As_2Se_3, but PbS-rich glasses resembled PbS itself, also in keeping with the idea of the equivalence of short-range order in amorphous and crystalline materials of the same structure. In the more complex glasses, however, this question becomes academic, since homogeneous crystalline solids of the same composition in general do not exist. The lesson of GeTe is that when the short-range order of a similarly constituted crystal and glass differ, the amorphous material appears to be the more covalent. Because of this and the fact that a covalently bound material is very resistant to crystallization, it is reasonable as a first hypothesis to assume that the multicomponent chalcogenide glasses are essentially covalent in nature and that all bonding requirements of each constituent atom is locally satisfied. In a complex material such as $Si_{18}As_{27}Se_{16}Te_{39}$, it is extremely difficult to envision a homogeneous crystal in which no dangling bonds exist, but it is not hard to construct a glass, given the additional freedom of positional disorder. Thus, each Si atom can find a tetrahedral position, with a local coordination number of 4, to form covalent bonds with all its outer electrons and satisfy its valence requirements. Similarly, each As atom can find a position with three nearest neighbors, in accordance with the 8-N rule of covalent bonding,[423] and each Se or Te atom can find a position with two nearest neighbors. If a halogen atom is present, only one nearest neighbor is required. There is some evidence in As-S-I glasses that single As-I bonds are indeed formed.[609] But only in a glass, with long-range positional disorder, can all covalent bonding requirements be satisfied. Such a structure may have microscopic pores and thus a density deficiency, but it will not have dangling bonds.

It is extremely difficult to confirm or reject such a hypothesis as this by physical studies. In principle, it is possible to observe dangling bonds by EPR experiments, although this has not proved to be useful in amorphous As_2Se_3. Agarwal and Fritzsche[482,483] observed a spin density of only 10^{15} - 10^{16} cm^{-3} in chalcogenide glasses, depending on the thermal history of the sample. This cannot be taken as evidence for the absence of a significant density of dangling bonds, however, since the relaxation time may be too fast to observe an EPR signal. In addition, the structural model can still be correct and there can still be a very large free-spin density if a significant CFO-type overlap of the valence and conduction band tails exists. In the CFO model, the redistribution of electrons between the valence and conduction bands that must occur to bring about thermal equilibrium must concomitantly create many dangling bonds.

A three-dimensional completely covalently bound glass might be expected to exhibit a high glass transition temperature. This has indeed been found to be the case, and materials such as $As_2S_3Ge_2$ have glass transition temperatures in excess of 700K.[610] In addition, replacement of the weak Van der Waals bonds present in materials like Se and As_2Se_3 by covalent bonds in the multicomponent glasses should lead to sharp increases in microhardness and stability, in accordance with observations.[432,611-613]

Infrared absorption peaks can be correlated with covalent stretching and bending frequency even in complex glasses,[551,614] strong evidence for the random covalent model under discussion. Furthermore, Mossbauer measurements[615] of Te^{125} in amorphous $Te_{70}Cu_{25}An_5$ yield a quadrupole splitting essentially the same as in pure Te, indicating the basically covalent nature of the bonding in the glass. On the other hand, a nuclear magnetic resonance study[616] of amorphous and crystallized $Te_{81}Ge_{15}As_4$ showed a 0.1% chemical shift of the Te^{125} peak from that of crystalline Te and from the crystallized alloy. This is still consistent with local satisfaction of all value requirements, provided the atoms have a small ionicity.[539]

2. Electrical Properties

For a large class of multicomponent chalcogenide glasses, electrical conductivity increases exponentially with increasing temperature, in accordance with the ordinary semiconductor expression.

$$\sigma = \sigma_o e^{-E_A/kT}$$

with σ_o between 10^2 Ω^{-1} cm^{-1} and 10^4 Ω^{-1} cm^{-1} and E_A in the 0.3 to 1.0 eV range.[130,451,617-619] Often $\sigma(T)$ exhibits no break either at the glass transition temperature or at the melting point.[443,452] However, this is not a general result, and sharp increases in the apparent activation energy (the slope of a plot of ln σ vs. 1/kT) can occur at temperatures between the glass transition temperature and the melting point.[450,620-622] It is important to note that interpretation of the apparent activation energy in terms of a physical energy difference is incorrect when E_A is varying with temperature.[623] For example, if the activation energy is decreasing quadratically with increasing temperature, then $E_A = E_o - A(kT)^2$, and

$$E_A = E_o - A(kT)^2 \quad ,$$

$$\sigma = \sigma_o \exp(-E_o/kT) \exp(AkT) \quad .$$

The apparent activation energy, Δ is then

$$\Delta \equiv - d \ln \sigma / d (1/kT)$$

$$= E_o + A(kT)^2$$

Thus, the apparent activation energy differs from the real value by $2A(kT)^2$. But, more important, the apparent activation energy is an *increasing* function of temperature, while the real activation energy *decreases* with increasing temperature. In Te_{48} As_{30} Si_{12} Ge_{10}, Δ increases from 0.05 eV at room temperature to 1.6 eV at 850K.[622] In accordance with the above discussion, this indicates a *decrease* in the separation between the Fermi energy and the valence band mobility edge.

Fagen and Fritzsche[624] measured the conductivity of thin films of amorphous Te_{40} As_{35} Ge_{11} Si_{11} P_3, and found that the as-deposited material exhibited an activation energy of 0.46 eV from 150K to 400K. At lower temperatures, some curvature of the $\ln \sigma$ vs. T^{-1} plot becomes evident. As the film was annealed at temperatures in the 400 to 450 K range, and low-temperature conductivity increased several orders of magnitude and the apparent activation energy in this region decreased, as shown in Figure 30. The process saturated at about 450 K (curve J), until the temperature was increased to 494K, at which point the conductivity sharply increased to about 0.3 Ω^{-1} cm^{-1} and the activation energy decreased to less than 0.01 eV. The last change clearly represents a cooperative crystallization of the material, perhaps with an accompanying phase separation. It is not yet clear what is responsible for the initial effects of annealing, although partial devitrification is a possibility. Croitoru et al.[625,626] also found a decrease in apparent activation energy of amorphous Te_{50} As_{30} Si_{11} Ge_9 at low temperatures, but they did not study the effects of annealing. Uphoff and Healy[627] determined that the activation energies in the As-Tl-Te-Se system depend critically on composition, and suggested that the apparent intrinsic-extrinsic transition as the temperature decreases in certain samples is due to phase separation. The sensitivity of the activation energy on composition appears to depend on the particular system being investigated,[627,628] and for a large class of chalcogenide glasses E_A is essentially independent of composition.[629] Hilborn and Prasad[630] performed annealing studies of the amorphous As-Te-I system, and found results similar to those of Fagen and Fritzsche.[624] Hilborn and Prasad noted that the isothermal conductivity increases linearly and irreversibly with time, with the rate of increase a strong but non-monotonic function of temperature. Since they found that the initially amorphous samples exhibited larger discontinuities upon annealing than samples that were originally partly devitrified, Hilborn and Prasad suggested the anomalous behavior of the conductivity was due to devitrification. Note that the electrical effects of annealing chalcogenide-glass films are the reverse of the situation or amorphous Ge and Si (see subsection IV A), for which the isothermal conductivity *decreases* with time after annealing.

Fagen et al.[631] measured the pressure dependence of the electrical conductivity of amorphous As_{35} Te_{28} Si_{21} Ge_{16} and Te_{80} Si_{20}, which have activation energies of 0.55 eV and 0.50 eV, respectively. If σ_0 is assumed to be independent of pressure, then for the quaternary alloy, $(d E_A/dp)_T = -1.0 \times 10^{-5}$ eV/bar at 300K, while for the binary alloy, $(d E_A/dp)_T = -1.4 \times 10^{-5}$ eV/bar. These values are of the same order of magnitude as the pressure coefficients of typical crystalline semiconductors.

Thermoelectric-power measurements have been performed on several of the multicomponent chalcogenide glasses.[442,580,627,632-634] In amorphous Te_{70} Cu_{25} Au_{51}, Tsuei[632] found a relatively sharp change in activation energy at about 180K, the low-temperature value being 0.05 eV, increasing to 0.11 eV at high temperatures. The thermoelectric power exhibited a maximum near 180K, in the vicinity of a discontinuity in the activation energy for electrical conduction. Although a thermoelectric-power maximum is an expected feature near a change from hopping to band-like conduction,[635] the conductivity below 180K does not exhibit an exp $(-A/T^{1/4})$ dependence and the thermoelectric power above 180K is not proportional to the logarithm of the resistivity, in disagreement with the predictions of a model based on a hopping-to-band transition. More typically,[442,627,633,634] the thermoelectric power decreases monotonically as a function of increasing temperature. In some cases,[634] the activation energy calculated from the thermoelectric data agrees well with that obtained from resistivity measurements, although often it is of the order of 0.1 to 0.3 eV lower.[627,634,636] In all cases reported for multicomponent chalcogenide glasses, the sign of the thermoelectric power indicated the predominance of p-type conduction.

The Hall effect has also been observed in chalcogenide glasses,[580,636-640] although the results are not at all consistent. Kolomiets and Nazarova[580] reported an n-type Hall voltage in the $Tl_2Se \cdot As_2(Se, Te)_3$ system. The Hall mobility was about 0.02 cm^2/V-sec, independent of composition. For all materials investigated, the thermoelectric power was observed to be p-type. Peck and Dewald[637] found similar anomalies in the As-Te-I and As-Te-Br systems. For amorphous As_{53} Te_{43} I_4, the Hall mobility was reported to

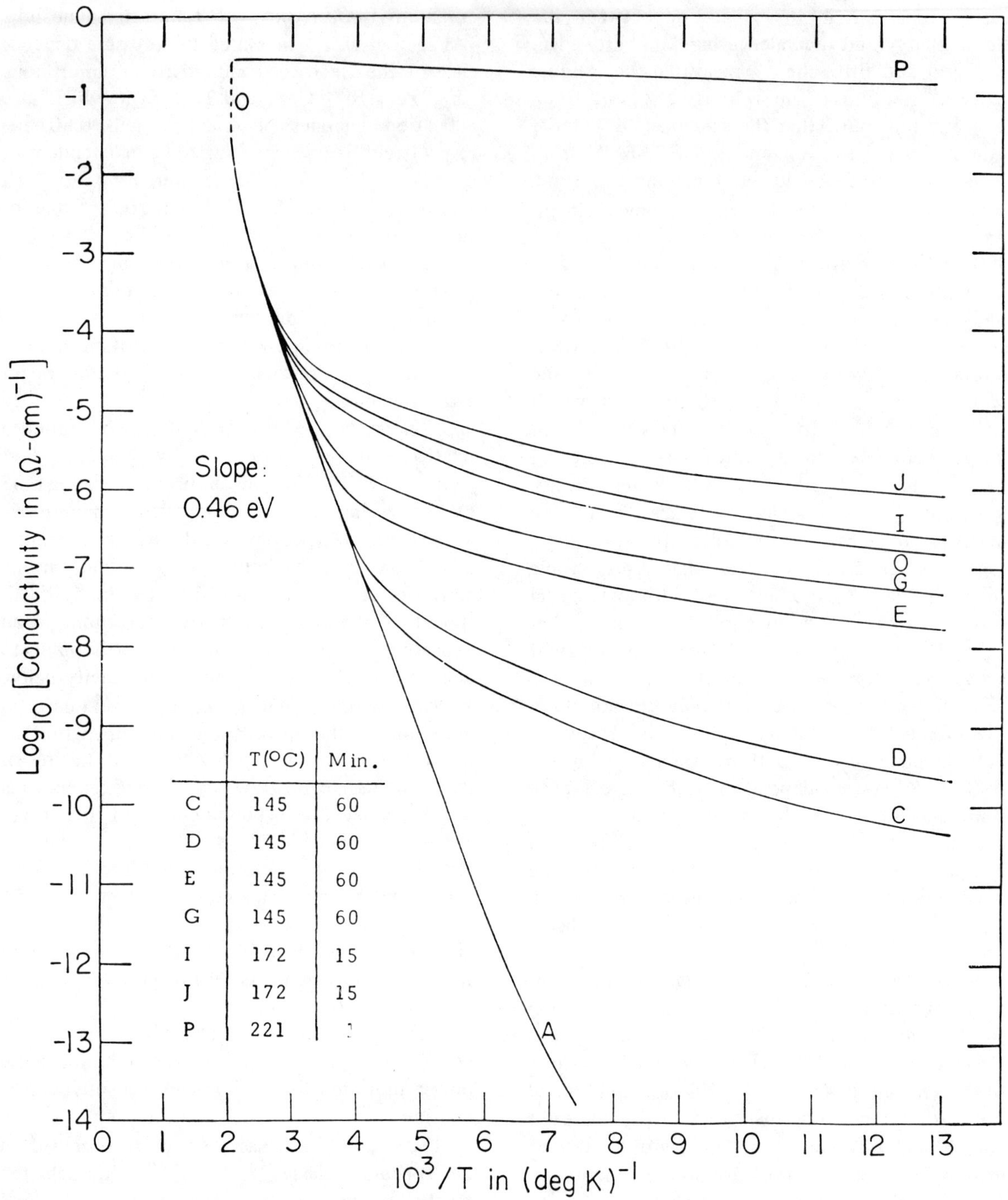

	T(°C)	Min.
C	145	60
D	145	60
E	145	60
G	145	60
I	172	15
J	172	15
P	221	1

FIGURE 30. Conductivity as a function of temperature for a thin film of amorphous $Te_{40}As_{35}Ge_{11}Si_{11}P_3$ subjected to thermal cycling procedure. Table (inset) gives maximum temperature previously attained, and duration of cycle (Fagen and Fritzsche[624]).

be 0.08 cm^2/V-sec at 300K, and it increased rather slowly with temperature to 0.12 cm^2/V-sec at 370K. The activation energy for conduction in this material was 0.4 eV, but the activation energy of the Hall mobility was only 0.03 eV. Thus, if the Hall constant can be interpreted in the usual manner, the activation energy can be almost entirely associated with the carrier concentration. Male[638] also observed an n-type Hall mobility for several glasses in the As-Se-Te and As-Se-Te-Tl systems, in which the thermoelectric power was always p-type. Some of these results are shown in Figure 31. The sign anomaly persists in the liquid phase, although upon crystallization, a rapid

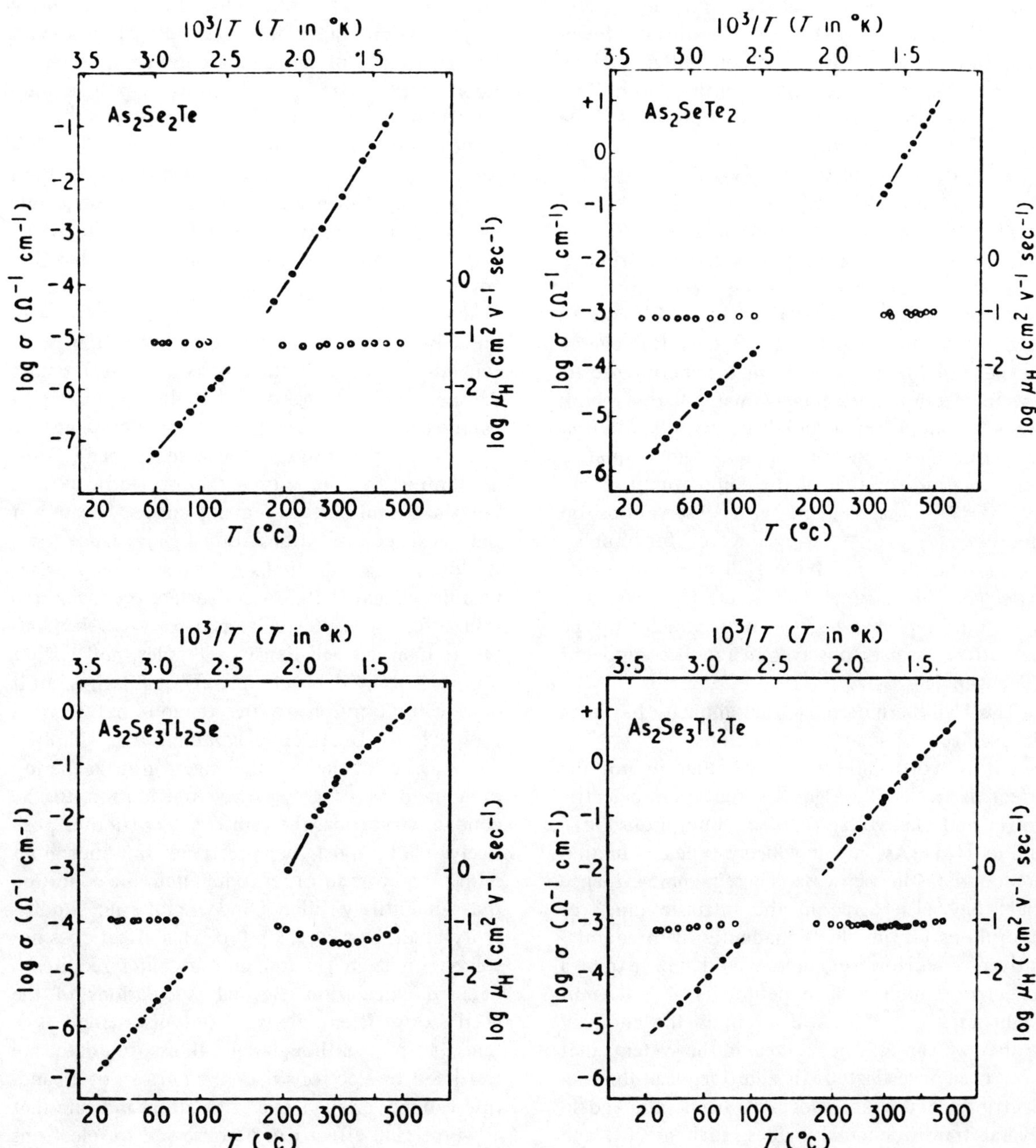

FIGURE 31. Hall mobility (open circles) and conductivity (closed circles) for chalcogenide semiconductors in the amorphous and liquid states (Male[638]).

change in the sign of the Hall voltage occurs. The Hall mobility is always in the vicinity of 0.1 cm^2/V-sec, and is independent of temperature. On the other hand, Nagels et al.[636] and Kornfel'd and Sochava[640] found a p-type Hall effect in amorphous $Tl_2Te \cdot As_2Te_3$, in agreement with the sign of the thermoelectric power. This material exhibits an activation energy for conduction of 0.25 eV. The activation energy deduced from the thermoelectric power is only 0.13 eV, significantly lower than that obtained from the conductivity. However, the activation energy resulting from Hall-effect measurements is 0.22 eV, essentially the same as that of the conductivity. The Hall mobility is of the order of 0.1 cm^2/V-sec, and shows a slight increase with temperature. Although the small activation energy (approximately 0.03 eV) associated with the Hall mobility can be attributed to a normal band-like scattering mechanism such as ionized-impurity scattering, it is difficult to account for the larger (0.12 eV) activation energy of the mobility deduced from the thermoelectric power. Interestingly enough, small-polaron theory predicts that the Hall mobility should exhibit an activation energy in the hopping regime equal to only one-third of that of the conductivity mobility,[641,642] in agreement with observations on amorphous $Tl_2Te \cdot As_2Te_3$. However, for hopping conduction, the sign of the Hall effect might be *expected* to be n-type;[641] in fact, $Tl_2Te \cdot As_2Te_3$ is the only reported case to date in which a p-type Hall effect has been observed in a multicomponent chalcogenide glass.

The Hall-thermoelectric sign anomaly has been the subject of many speculations,[125,643-647] and it is worthwhile asking whether or not the effect is real. The signs of the thermoelectric power and Hall voltage agree in amorphous GeTe and in $Tl_2Te \cdot As_2Te_3$, in which p-type conduction predominates in each case. There is some evidence that they also agree in the intrinsic range of amorphous Ge, in which conduction is predominantly n-type. However, there would appear to be a sufficient number of experiments in different countries[580,637,638] which obtain the anomaly so that we can assume it is real in the systems that have been investigated. In addition, Hall-thermoelectric sign discrepancies have been observed in several transition-metal oxides, such as NiO and Fe_2O_3,[648] over a wide temperature range, and in many liquids, such as Te.[649,650] Although specialized explanations have often been proposed,[649,651] the generality of the anomaly is staggering. However, it must be borne in mind when analyzing any proposed resolution that the anomaly is not universal, and we must explain why it exists, for example, in $Tl_2Se \cdot As_2Te$ but not in $Tl_2Te \cdot As_2Te_3$. Pearson[643] suggested the n-type Hall effect results from a crystalline phase precipitating in the glassy matrix, but this is difficult to reconcile with Male's[638] observation of a rapidly varying Hall mobility during crystallization or with the persistance of the anomaly in the liquid phase. Boer[646] proposed that electrons as well as holes contribute to both the Hall effect and the thermoelectric power, and suggested a model in which the sum of these contributions could exhibit opposite signs over a large temperature range, due to sharp increases of carrier mobility with energy above the mobility edges. This model has the advantage of easily accounting for the low values observed for Hall mobility, but it is extremely unlikely that it could account for the wide generality of the Hall-thermoelectric sign anomaly. Allgaier[644,645] suggested that the apparent sign paradox could be resolved if the Fermi energy were near the top of the valence band, thus accounting for the p-type thermoelectric power, but the Fermi surface remains convex because of the absence of translational invariance and Brillouin zone walls in disordered systems. Allgaier postulated that if the Fermi surface is convex, the Hall effect measures a large density of electrons rather than a small density of holes, and is thus n-type. This model is completely qualitative, but it does call attention to the tenuous nature of a p-type Hall effect. Hole *conduction* is simply a consequence of the fact that electrons near the top of a band make a negative contribution to the conductivity; thus, their absence contributes positively. This must be preserved in amorphous semiconductors in order to maintain the condition that an entirely filled band exhibits no conductivity at all. However, a p-type Hall effect does not follow on such general grounds, and requires a detailed calculation. Several calculations of the Hall coefficient exist for disordered systems,[652,653] although all those to date are restricted to electrons near the *bottom* of a band. Interestingly enough, the calculations indicate that a p-type Hall effect can be observed for electrons in a conduction band when the Fermi energy lies in the band tail below the mobility edge. Unfortunately, this is the reciprocal of the desired result, which is that an n-type Hall effect can be observed for holes in the valence band.

The ac conductivity of several chalcogenide glasses has been measured,[654-656] and found to be quite dependent on the composition being investigated. Bishop et al.[655] studied amorphous $Tl_2Se \cdot As_2Te_3$, a material in which the ac conductivity equals the dc conductivity for frequencies up to at least 10^{10} Hz at 300K. On the other hand, Rockstad[654] found that the ac conductivity of $Te_{48}As_{30}Si_{12}Ge_{10}$ and Te_2 As Si essentially equals the dc conductivity only up to 10^4 to 10^5 Hz, but then increases with frequency approximately proportional to $\omega^{0.9}$ in the 10^5 to 10^8 Hz region. This apparent discrepancy is easily resolved by noting that $Tl_2Se \cdot As_2Te_3$ is a glass of unusually high dc conductivity, about 3 x 10^{-3} Ω^{-1} cm^{-1} at 300K, and this undoubtedly dominates any hopping component at that temperature. The two glasses investigated by Rockstad have dc conductivities less than 10^{-5} Ω^{-1} cm^{-1}, and thus the ac component predominates at a lower frequency. The ac conductivity results for amorphous Te_2AsSi are shown in Figure 32. Rockstad[656] measured the temperature dependence of the ac conductivity of amorphous $Te_{48}As_{30}Si_{12}Ge_{10}$ and found that below 200K, the frequency dependent component is approximately proportional to T, as Equation 29 predicts. This is strong evidence that this component is due to hopping in the vicinity of the Fermi energy. Use of Equation 29 gives a density of states at the Fermi level of about 3 x 10^{18} cm^{-3} eV^{-1}. Rockstad[656] also was able to separate out a third component of the ac conductivity. This component is thermally activated, but with an activation energy lower than that of the dc conductivity. The conductivity of this third component also increases approximately proportional to ω, and it can thus be associated with hopping either within one of the band tails or in an impurity band.

Fritzsche, et al.[657,658] performed field-effect measurements on a multicomponent chalcogenide glass in order to directly measure $g(E_f)$ by using Equation 26. The density of injected electrons, ΔN, was determined from capacitance and voltage measurements. It was hoped to calculate ΔE_f from direct measurements of the change in conductivity, $\Delta\sigma$. If the predominant carriers are holes whose mobility is unaffected by the injection, then

$$\frac{\Delta\sigma}{\sigma} = \frac{\Delta p}{p} = e^{-\Delta E_f/kT} - 1$$

$$\approx -\frac{\Delta E_f}{kT} ,$$

since $p = p_o e - (E_F - E_v)/kT$. Thus, ΔE_F can be

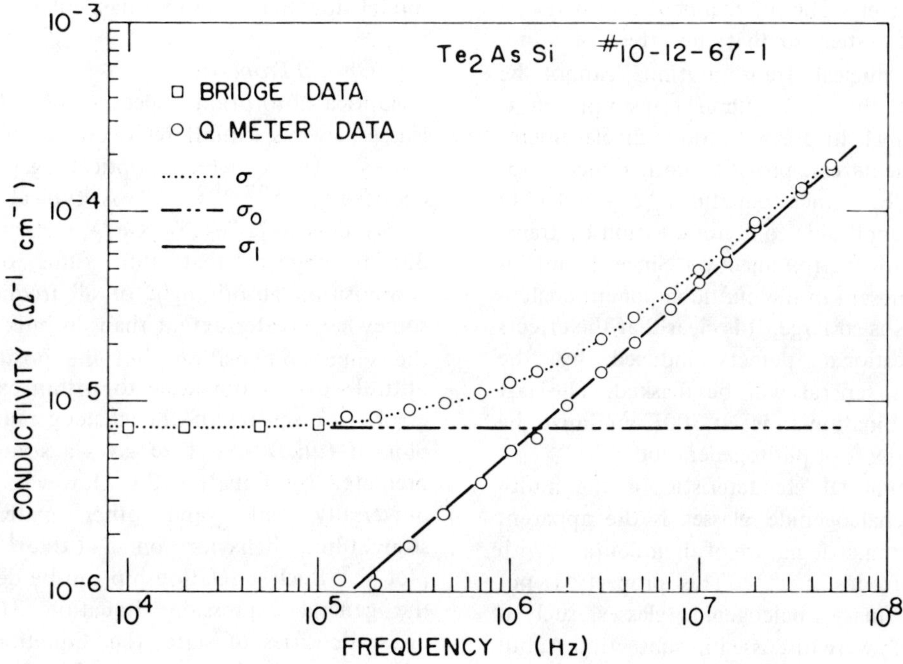

FIGURE 32. Conductivity as a function of frequency for amorphous $Te_{48}As_{30}Si_{12}Ge_{10}$ (Rockstad[654]).

determined, and Equation 26 gives $g(E_F)$. Unfortunately, for the injection density used, no detectable conductivity change was observed. Thus, only a lower limit, $g(E_f) < 5 \times 10^{19}$ cm^{-3} eV^{-1}, was obtained. This value appears to be rather on the large side. Barbe[659] calculated the change in conductance of a particular field-effect geometry as a function of gate voltage, taking into account the effect of surface states and band bending. He found that if the semiconducting film is sufficiently thin, then conductance changes should be measureable, provided that the surface-state density does not exceed about 10^{14} cm^{-2} eV^{-1}, a value larger than usual for amorphous films.[660]

Tick and Hindley[661] recently reported a field effect in a multicomponent chalcogenide glass, using a concentric cylindrical geometry. They found a surface-state density of 6×10^{13} cm^{-2} eV^{-1}, and estimated $g(E_f) \sim 3 \times 10^{16}$ cm^{-3} eV^{-1} The large discrepancy between this value and that obtained by Fritzsche et al.[657] is not at all clear.

Ovshinsky et al.[658] investigated the radiation hardness of devices based on amorphous $Te_{48}As_{30}Si_{12}Ge_{10}$ by exposing them to flash x-ray pulses up to 1.8×10^{11} rads/sec and to fast-neutron fluences up to 1.2×10^{17} n/cm^2. The devices continued to function, although deterioration of the cables led to up to a 30% variation in device parameters. The thermal-neutron fluence, if any, was not stated, so that the effect of donor creation by nuclear transmutations cannot be estimated. If the recoil upon transmutation is sufficiently small to prevent atomic displacement, such experiments can provide another means for estimating $g(E_f)$, since Equations (26) and (41) should be as applicable to donor creation by transmutations as to electron injection. Since the intrinsic disorder present in the multicomponent chalcogenide glasses is so large, it is clear that the effects of the additional defects induced by the irradiation in general will be masked. The fast carrier recombination times at 300K minimize the deleterious effects of photogeneration.

Another unusual characteristic of the multicomponent chalcogenide glasses is the apparent ohmic and symmetric nature of their contacts with metallic electrodes.[618,658] This property is not observed in binary chalcogenide glasses, such as As_2Se_3, which were discussed in subsection A, but it appears to be virtually universal in the complex glasses. The ohmic nature of the contacts could simply be a masking of contact resistance by the high resistance of the material itself, but little information bearing on this point is available at present. If the effect is real, however, two possible explanations are available, depending on the as yet undetermined value of $g(E_f)$. Ovshinsky et al.[658] assumed $g(E_f)$ is of the order of 10^{20} cm^{-3} eV^{-1}, in accordance with the field-effect results of Fritzsche et al.,[657] and pointed out that the corresponding trap density is then sufficiently large to accommodate the space charge near each electrode within a 30 Å region. Such narrow Schottky barriers would always allow carrier injection to freely occur by means of tunneling, thus yielding ohmic contacts. Fritzsche[483] recently proposed a model that explicitly considers the spatial potential fluctuations resulting from the positional and compositional disorder. Since electron and hole traps are spatially separated, the material can be looked at as extremely unhomogeneous, and being made up of intertwined p-rich and n-rich regions. Since any cross section of the material of 1μ or greater extent contains large numbers of both types of region, any metallic electrode will simultaneously make both ohmic and blocking contact with parts of the glass. The ohmic regions of the junctions will then provide low-resistance contacts, in accordance with the observation. Fritzsche's model does not require a particularly large $g(E_f)$.

3. *Optical Properties*

Optical-absorption spectra of chalcogenide glasses always exhibit relatively sharp absorption edges, from which optical gaps can be deduced.[579,607,662–666] A typical result, for amorphous $As_{35}Te_{28}S_{21}Ge_{16}$, is shown in Figure 33. It appears that thin films of a given composition absorb light of all frequencies to a somewhat greater extent than do bulk samples of the same composition, but the position of the optical edge is the same to within experimental error. For several of the chalcogenide glasses, a plot of $(\alpha\hbar\omega)^{1/2}$ vs. $\hbar\omega$ gives a straight line, as predicted by Equation 34. However, this is not universally valid, and other materials show straight-line behavior on a $(\alpha\hbar\omega)^{1/3}$ vs. $\hbar\omega$ plot.[667] Such a relationship can be derived from the general expression, Equation 31, provided linear densities of states (i.e. Equation 33) exist near the band edges and provided the absorption matrix elements are essentially independent of

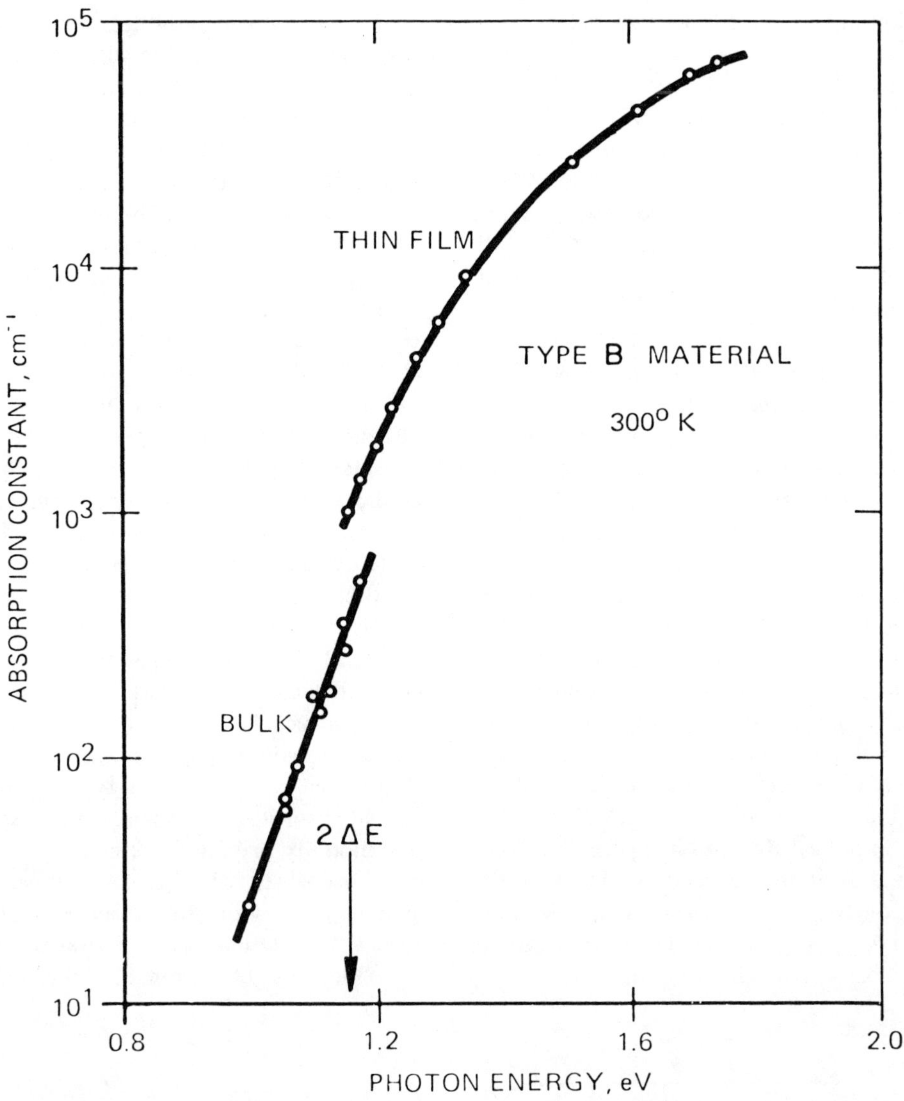

FIGURE 33. Optical absorption coefficient as a function of photon energy for bulk and thin-film samples of amorphous As_{35} Te_{28} S_{21} Ge_{16} (Fagen and Fritzsche[666]).

energy throughout the entire range.[130] Although it would appear that both these restrictions are unlikely to be appropriate simultaneously, they do give agreement with experiment for a class of materials. The optical gaps obtained for multicomponent chalcogenide glasses are always approximately twice the activation energy for electrical conduction, a surprising result in view of the apparent pinning of the Fermi energy. There is no obvious reason why, in either a CFO model or a compensating trap model, the Fermi energy should always be near the center of the mobility gap, although such a result can emerge from the CFO approach if the valence and conduction bands tail symmetrically.

The temperature coefficient of the optical edge of chalcogenide glasses is the same order of magnitude as in representative crystalline semiconductors, with observed values lying between -4 x 10^{-4} eV/K and -10 x 10^{-4} eV/K.[452,631,666] The pressure coefficients of the optical gaps are also similar to those of crystalline semiconductors, and have been observed in the range -1 x 10^{-5} eV/bar to -2 x 10^{-5} eV/bar.[631] However, the optical gap appears to decrease with pressure at a slower rate than does $2E_A$, indicating that the close relationship between the two values may indeed be accidental.

Fagen and Fritzsche[666] studied electro-absorption of thin films of amorphous $As_{35}Te_{28}S_{21}Ge_{16}$ at 77K, and found a small shift, of the order of 3×10^{-15} eV-cm^2/V^2, comparable to the room-temperature value observed in amorphous Se,[401-403] but approximately a factor of 100 smaller than that for crystalline Se[401] or CdS.[668,669] The electro-absorption signal appears to increase with increasing temperature, in disagreement with the predictions of the Franz-Keldysh effect,[405] but Fagen and Fritzsche point out that at room temperature a small thermoabsorption modulation induced by Joule heating could reverse the observed temperature dependence of the signal.

Fritzsche[483] has focused on the problem of the apparent discrepancy in values of $g(E_f)$ calculated from electrical and optical measurements on the same materials. As previously indicated, electrical experiments themselves are not yet in agreement on the order of magnitude of $g(E_f)$, but for both binary and multicomponent chalcogenide glasses, the ac-conductivity results appear to yield values of $g(E_f)$ of the order of 10^{18} cm^{-3} eV^{-1}. On the other hand, in optical-absorption experiments, the value of the absorption coefficient typically drops by a factor of 10^4 in approximately 0.5 eV. For photon energies corresponding to the activation energy for electrical conduction, the absorption coefficient is generally less than 10 cm^{-1} and essentially unmeasurable. But if the argument presented in section III that the matrix elements for delocalized-to-localized optical transitions are of the same order as those for transitions across the mobility gap is correct, then we can conclude that $g(E_f)$ is considerably less than 10^{17} cm^{-3} eV^{-1}. Fritzsche pointed out the fact that optical absorption can be very insensitive to spatial potential fluctuations, as, for example, in a crystalline p-n junction. Although it appears from the conventional sketch of the spatial dependence of the energy band structure of a p-n junction (Figure 34) that photons with energies considerably smaller than the actual band gap can connect band-like states, these photons, in fact, are not absorbed by the junction. Thus such a sketch, although very useful for discussing electrical properties, can be misleading with respect to analyzing optical properties (see also ref. 122). Fritzsche suggested that similar effects hold throughout amorphous semiconductors, which, since electrons and holes are trapped in spatially-separated regions, can be thought of as alternating p-type and n-type material. Thus, optical absorption between localized states deep in the mobility gap and extended states near the mobility edges will be strongly suppressed, in disagreement with the calculation of Davis and Mott.[130] Fritzsche's model is sketched in Figure 35. One way of reconciling the approaches of

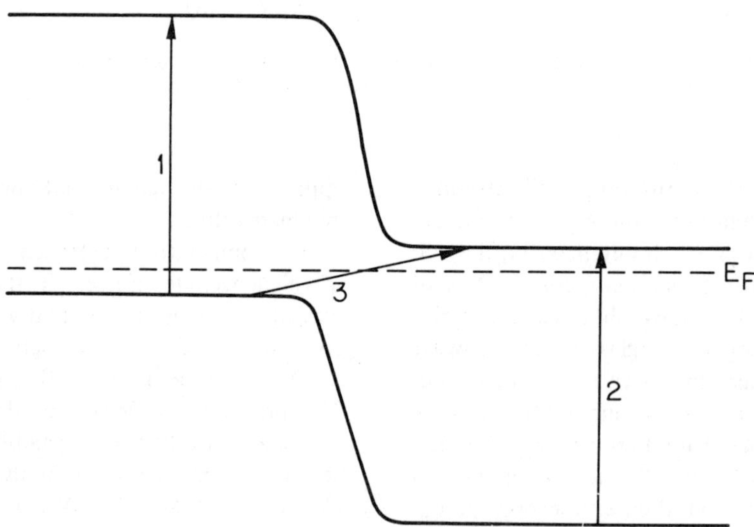

FIGURE 34. Energy band structure for an ordinary crystalline semiconductor p-n junction. The transition labeled 3 cannot be excited optically.

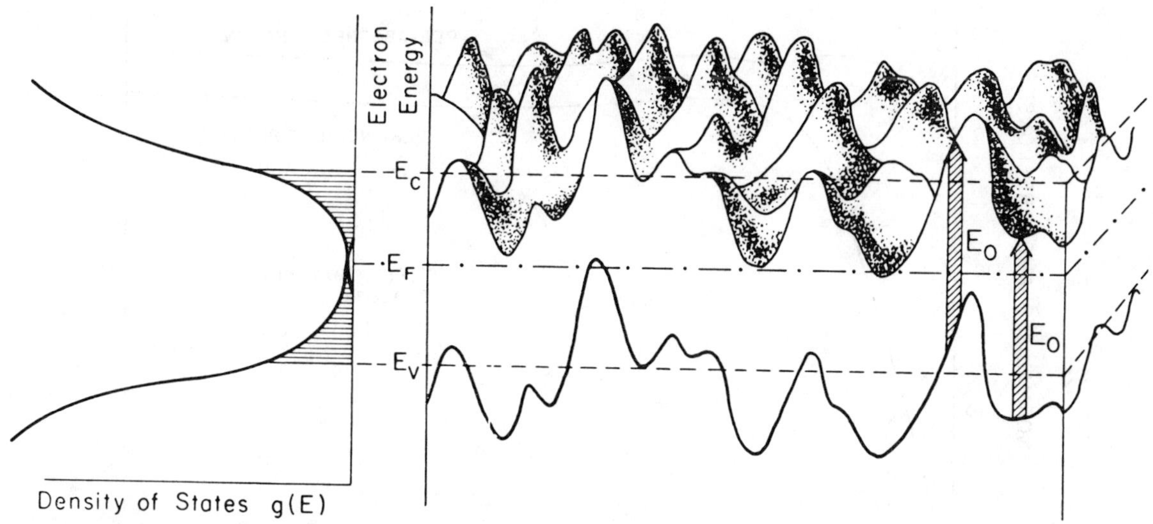

FIGURE 35. Potential fluctuations of the initial and final electron states for the optical transitions corresponding to the optical gap E_o. The left hand side shows the density of states. The region of localized states lies between E_c and E_v. Note that the short range potential wells which give rise to many of the localized states are not shown here. This figure shows only that part of the long wavelength potential fluctuations which cause a parallel shift of the valence and conduction band states. The part which causes a spatial variation of E_o is omitted for clarity (Fritzsche[483]).

Fritzsche and of Davis and Mott is to note that in the former model the delocalized wave functions near, for example, the conduction-band mobility edge are excluded from the vicinity in which localized electrons are trapped. This effect serves to sharply reduce the corresponding matrix element for optical absorption. But this wave-function exclusion is, of course, a correlation effect, and thus could not be derived from the one-electron approach used by Davis and Mott. In connection with this, it is interesting that the simple model for introducing electronic correlations into the CFO model discussed in section III serves to produce a sharp optical absorption edge, as can be seen from Figure 9.

Photoconductivity has been commonly observed in chacogenide glasses,[500,626,666,670-674] and the major results appear to be essentially independent of compositon. Holes rather than electrons always appear to dominate the photocurrent,[452] in agreement with the thermoelectric results. Fagen and Fritzsche[672] studied the photoresponse of amorphous $Te_{40}As_{35}Si_{11}Ge_{11}P_3$ in great detail. This material has an activation energy for electrical conduction of 0.45 eV and an optical gap near 0.90 eV.[624,666] The normalized photoresponse as a function of photon energy is shown in Figure 36. Since the quantum efficiency is essentially uniform from 0.8 eV through 3.0 eV, it is almost undoubtedly the case that it is numerically equal to unity. This result is common to many chalcogenide glasses. No non-photoconductive absorption, as is found in amorphous Se, has been observed in the complex glasses. At 300K, the photocurrent is proportional to the light intensity, while at 77K, the photocurrent is proportional to the square root of the intensity. Thus, it can be concluded that bimolecular recombination processes predominate at low temperatures, but monomolecular processes predominate at high temperatures. However, at intermediate temperatures, the intensity dependence of the photocurrent is intermediate between square root and linear, indicative of a continuous distribution of traps in the gap.[675] This result is contrary to the results on amorphous As_2Se_3, in which the transition from bimolecular to monomolecular recombination is rather abrupt.[497]

Fagen and Fritzsche[672] observed a maximum in the photoresponse near 250K, as is shown in Figure 37. Such a maximum has been observed in several other chalcogenide glasses,[497,671,676] and represents a maximum in the product of carrier mobility and lifetime. The maximum could represent the point at which equilibrium conduction begins to dominate photoconduction, since at higher temperatures the lifetime will begin

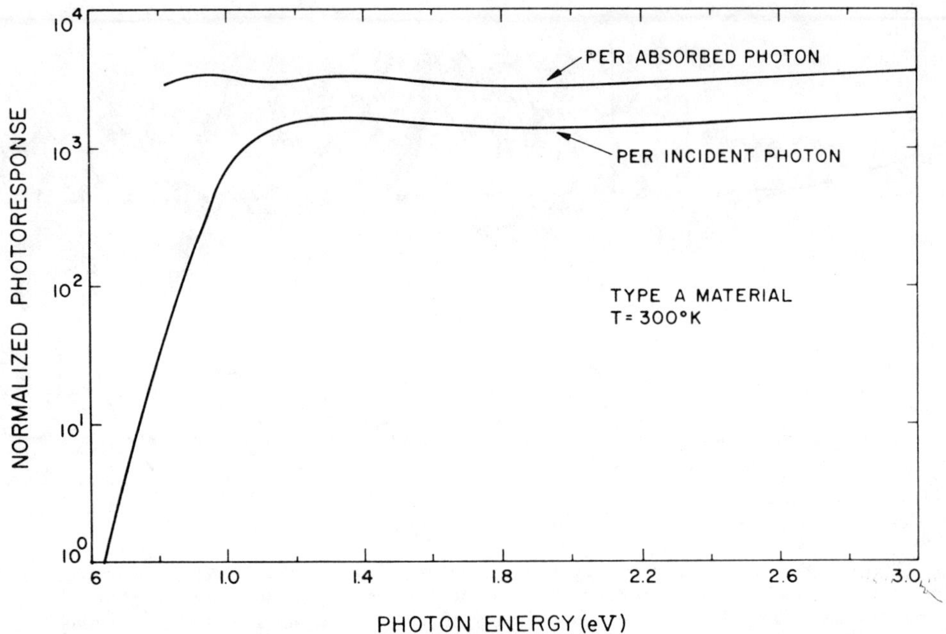

FIGURE 36. Normalized photoresponse as a function of photon energy for a thin film of amorphous Te_{40} As_{35} Si_{11} Ge_{11} P_3 at 300 K. The upper curve is relative quantum efficiency (Fagen and Fritzsche[672]).

to decrease proportional to the reciprocal square of the intrinsic carrier concentration.[671] The thermally activated behavior at temperatures below the maximum indicate that traps centered about 0.2 eV above the valence-band mobility edge predominate. Similar results, with the same activation energy, have been found in $Te_{50}As_{30}Si_{11}Ge_9$,[671] in GeTe,[573] and in $As_2Se_3 \cdot 2AsTe$.[513] In As_2Se_3, which has a larger mobility gap, the corresponding value is 0.35 eV,[499] while in both As_2Te_3 and Te_2AsSi, which have smaller mobility gaps, a 0.15 eV activation energy of the photoresponse has been observed.[626]

Studies of. the decay kinetics of the photoconductivity have been extremely fruitful in the chalcogenide glasses. Even at room temperature, two distinct relaxation times can be identified after the illumination is turned off,[671] but the effect is particularly striking at 77K, as is evident from Figure 38. Most of the photocurrent decays relatively quickly, but approximately 10% persists for hours, resulting in an excess dark conductivity several orders of magnitude larger than the equilibrium value. Such a result is far from new, and has been observed, for example, in amorphous As_2Se_3[499] and $Tl_2Se \cdot As_2Te_3$[670], in heavily doped crystalline Ge,[677,678] Si[679] and GaAs,[680,681] in crystalline Se[682] and MnO,[683] and in irradiated Si.[684] The explanation is always essentially the same, and depends on the existence of a substantial density of charged trapping levels deep in the gap. The possible recombination processes for photogenerated electrons are indicated in Figure 39. Direct recombination of mobile electrons and holes across the mobility gap, process (a), has a large cross section and must be an important mode of decay. Process (b), recombination of free electron with localized holes trapped below the Fermi energy has a somewhat smaller cross section, and is unlikely to dominate direct recombinations. But if a sufficiently large density of positively charged electron traps exists above the Fermi energy, process (c) can be the predominant decay mode, despite the smaller cross section compared to that of direct recombination. However, process (c) results in an electron being trapped above the Fermi energy, a non-equilibrium situation. Once trapped, the electron is highly localized and can only escape with the assistance of phonons. At room temperature, a sufficient concentration of phonons exists so that equilibration quickly occurs.

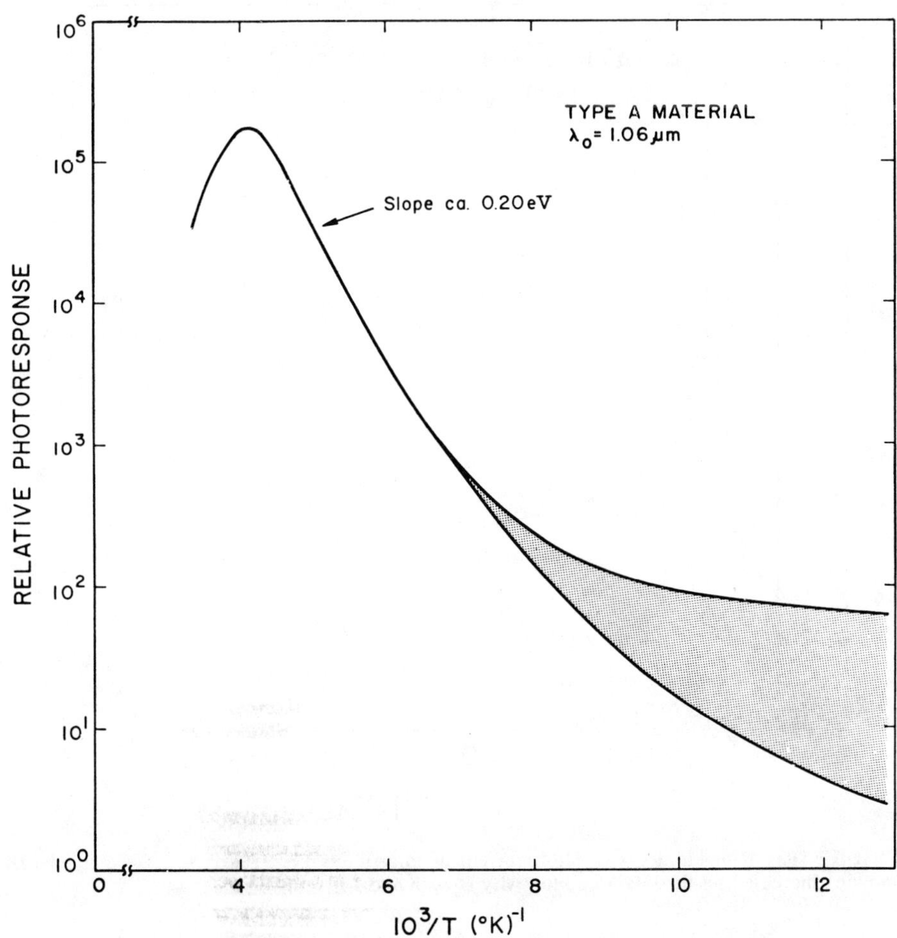

FIGURE 37. Resistive photoresponse as a function of temperature for a thin film of amorphous Te_{40} Ap_{35} Si_{11} Ge_{11} P_3. The shaded area indicates the range of sample-to-sample variation (Fagen and Fritzsche[672]).

However, at 77K, exponentially fewer phonons are present, and the electron can stay trapped for hours, resulting in an excess dark conductivity, as is observed. A large density of positively charged electron traps above the Fermi energy and negatively charged hole traps below the Fermi energy is an immediate consequence of the CFO model, but such traps also result from compensating donor and acceptor levels. Thus, once again, the difficulty of differentiating between these two models is manifest.

Botila and Vancu[671] found that the slow component of the photoconductive decay of amorphous $As_{50}Te_{30}Ge_{11}Si_9$ was thermally activated, with an activation energy of 0.2 eV, precisely the same as the activation energy associated with the photocurrent, as would be expected. This is some evidence in favor of the compensated-semiconductor model, since the activation energy for conduction is considerably larger than 0.2 eV.

Fagen and Fritzsche[672] studied thermally stimulated conductivity in amorphous $Te_{40}As_{35}Si_{11}Ge_{11}P_3$. The thermally stimulated current shows a peak near 210K, which yields[507] a trap energy of approximately 0.43 eV, essentially equal to the activation energy for electrical conduction. This suggests that the majority of the charged traps in this material lie near the Fermi energy, evidence in favor of the CFO model. Fagen and Fritzsche[672] further approximated an upper bound to the trap density from the area under the TSC curve,[685] a method which leads to a trap density of about 4 x 10^{19} cm^{-3}. Fagen[673] has noted that the fast component of the photoconductive decay appears

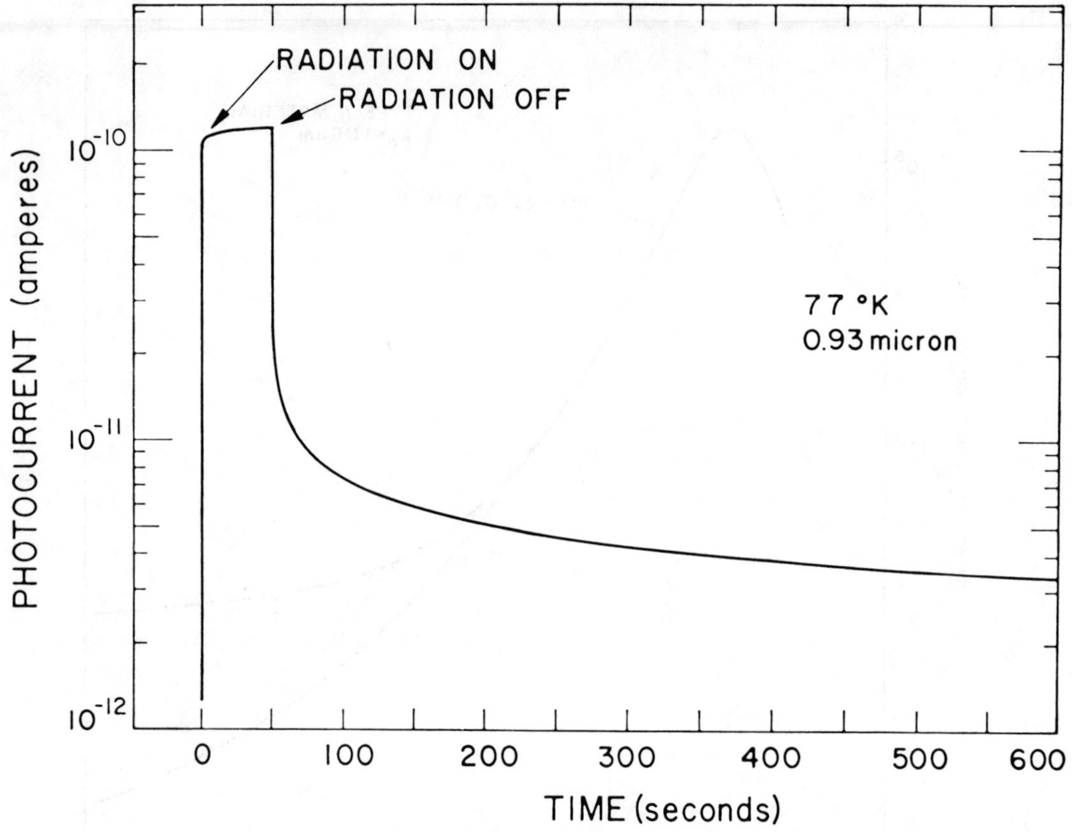

FIGURE 38. Rise and decay of photocurrent in amorphous Te_{40} Ao_{35} Si_{11} Ge_{11} P_3 at 77K, showing increased residual dark conductivity (Fagen and Fritzche[666]).

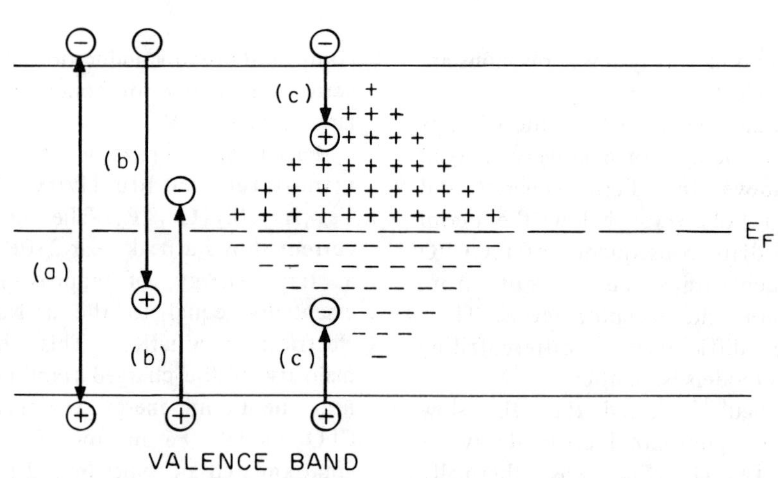

FIGURE 39. Possible recombination processes within the CFO model.

to be thermally activated at high temperatures, the activation energy being about 0.45 eV, in agreement with the TSC result. Finally, Fagen and Fritzsche[666] observed an excess conduction at 77K after exposing an amorphous film of $Te_{40}As_{35}Si_{11}Ge_{11}P_3$ to an increased electric field for several seconds. The slow component of the decay persisted for hours, completely analogous to the photoconductivity result.

4. Conclusions

Despite their extremely complex nature, great studies have been made in the past few years in understanding the multicomponent chalcogenide glasses. These materials intrinsically possess such a large degree of disorder, compositional as well as positional, that theoretical analysis based on first principles are clearly impossible. Nevertheless, thorough experimental investigations have been performed, and have shown that these easily fabricated glasses possess several unique and desirable physical properties. Although these properties were by and large totally unexpected, it is interesting that most of them can be explained rather simply from a number of widely divergent models.

At the present state of our knowledge, the following conclusions can be drawn.

1. The structure of these glasses is far from a closed question, but a random covalent model is consistent with the available data. Evidence from NMR experiments is that the constituent atoms have some degree of ionicity, but the vast majority of covalent valence requirements appear to be locally satisfied, and this is responsible for the metastability of the glasses.

2. The activation energy for electrical conduction is generally constant over many decades. There is some curvature in a plot of log σ vs. T^{-1} at low temperature, but no $T^{-1/4}$ behavior has been observed. In contrast to the situation for amorphous Ge and Si, annealing chalcogenide glasses below their crystallization temperatures irreversibly increases the electrical conductivity. This is perhaps due to phase separation.

3. Thermoelectric measurements indicate that the predominant carriers are always holes, a conclusion that is confirmed by photoelectric experiments. The predominance of hole conduction is a general result which thus far is unexplained. No drift mobility or photoinjection experiments have been reported as yet, but indication from observations on some of the simpler chalcogenide glasses are that the p-type behavior results primarily from a hole mobility that is larger than the electron mobility, rather than from the Fermi energy being nearer the valence-band mobility edge. The temperature dependence of the thermoelectric power suggests that the mobility increases with increasing temperature.

4. The Hall mobility is essentially independent of temperature. It is usually n-type, despite the apparent predominance of hole conduction. This sign anomaly has not yet been resolved, although it is a very general characteristic, in both amorphous and liquid semiconductors.

5. Evidence for hopping at the Fermi energy comes from ac-conductivity experiments, which show an increase in conductivity approximately proportional to both frequency and temperature at high frequencies and low temperatures. Another contribution to the ac conductivity, hopping among trap-like levels in the mobility gap, but somewhat removed from the Fermi energy, has also been observed. This is further evidence for the existence of well-defined trapping levels in chalcogenide glasses. These traps are probably due to structural defects in particular films, rather than due to any intrinsic effect.

6. The optical gap in all materials thus far investigated is approximately twice the activation energy for electrical conduction. This indicates that the Fermi energy is always near the center of the mobility gap, an unexplained result. The pressure coefficients of the two quantities appear to be quite different, however.

7. No non-photoconductive absorption has been observed in chalcogenide glasses, indicative to the absence of excitonic transitions. The quantum efficiency appears to be equal to unity over a wide range of photon energies.

8. The photoconductivity exhibits an activation energy, usually of the order of one third to one half of the activation energy for electrical conduction, although in several materials the two energies are almost equal. The same activation energy as is found in the photoconductivity often also appears in the photoconductive decay times. This is strong evidence for the existence of a large density of negatively charged hole traps in the mobility gap. In some materials, these traps appear to lie primarily near the Fermi energy, in accordance with the CFO model, but in others,

they appear to be considerably nearer the valence-band mobility edge. There is thus good reason to believe that the former are *intrinsic* traps, while the latter are *extrinsic* traps, their density depending on purity and method of preparation.

9.The photoconductive decay consists of a fast and a slow component. The latter becomes exponentially long at low temperatures, resulting in an excess dark conduction that persists for hours at 77K. This is further evidence for the existence of a large density of negatively charged hole traps. Thermally-stimulated-conductivity data indicate that the trap density is of the order of 10^{19} cm^{-3}.

10.The Fermi energy is somewhat pinned, but the order of magnitude of $g(E_f)$ is not clear as yet. The ac conductivity indicates that $g(E_f)$ is of the order of 10^{18} cm^{-3} eV^{-1}, but field effect experiments have given values ranging from 3×10^{16} cm^{-3} eV^{-1} to greater than 5×10^{19} cm^{-3} eV^{-1}. As Fritzsche has observed, correlation effects connected with the trapping of electrons and holes prevent the optical spectrum from being a useful method for estimating $g(E_f)$.

VI. IONIC AMORPHOUS SEMICONDUCTORS

Since structural rearrangements are considerably less difficult in ionic materials than in covalently bound materials, it might be expected that few primarily ionic glasses exist. This is actually not the case, and a surprisingly large number of such materials have been prepared. Vanadate and tellurite based glasses have been known for over a century,[686-689] and more recently, aluminate,[690-693] carbonate,[694] titanate,[695,696] sulfate,[697] nitrate,[698-700] Ga_2O_3,[701] MoO_3,[702] WO_3,[703] fluoride[704,705] and $ZnCl_2$ [706,707] glasses have been studied. More modern techniques, such as vapor deposition and rf sputtering, have led to the preparation of amorphous SnO_2,[708] ZnO,[709] NiO,[710] V_2O_5,[711] Cr_2O_3,[712] and Al_2O_3,[713] among many others. Electrical properties were studied extensively,[714] and it was generally concluded that conduction is primarily ionic rather than electronic. However, when the conductivity of vanadiumphosphate glasses was found to be large, of the order of 10^{-5} Ω^{-1} cm^{-1}, [715-718] and independent of time,[716,719] it became obvious that the material is an electronic conductor. This discovery opened up the field of semiconducting oxide glasses, and it was not long before the electrical properties of other vanadate glasses,[720-723] and glasses based on MnO, CoO, and iron oxides[724-728] were investigated in detail. Since we are primarily concerned with amorphous semiconductors rather than amorphous insulators in this review, we shall restrict our discussion to the transition-metal oxide glasses. Of these, by far the most thoroughly investigated class is the V_2O_5 - P_2O_5 system.

A. Vanadium Phosphate Glasses

1. Structure

X-ray, neutron, and electron diffraction data are generally unavailable in ionic glasses. In fact, the only oxide glasses for which radial distribution functions have been analyzed in any detail are those based on SiO_2,[729-732] B_2O_3,[733,734] P_2O_5,[735,736] TeO_2,[737,738] and As_2O_3,[739] all primarily covalently bound. For the primarily ionic oxide glasses, structural information must be extracted from infrared, EPR, and NMR data.

Crystalline V_2O_5 has a complex orthorhombic structure,[740] in which VO_5 pyramids link together partly at the corners and partly at the edges to form two-dimensional layers. Each VO_5 pyramid consists of a central V^{5+} ion, an O^{2-}ion as the apex, and four basal O^{2-} ions which are coplanar. Three crystalline allotropes of P_2O_5 exist,[741] a hexagonal, an orthorhombic, and a tetragonal form. All three structures are based on a PO_4 tetrahedron, in which a phosphorus atom is at the center and four oxygen atoms are at the corners. The hexagonal form consists of discrete P_4O_{10} molecules,[742] and is thus a molecular crystal held together by Van der Waals' interactions. The orthorhombic form is made up of rings of ten PO_4 tetrahedra,[743] the inter-ring forces again being of the Van der Waals type. Finally, the tetragonal allotrope is a layer structure, consisting of rings of six PO_4 tetrahedra.[744] Thus, none of the crystalline forms of P_2O_5 bear any resemblance to the V_2O_5 structure.

A major problem in trying to analyze the structure of vanadium phosphate glasses is to decide on the ionic state of the constituent atoms. It is usually assumed that only V^{5+} and V^{4+} ions are found in the glass, that the oxygen atoms surrounding a vanadium atom will gain an electron from it, but that the P-O bonds are essentially covalent. But even with these restrictions, the possibilities for short-range order are many fold.

One of the first techniques used to try to determine the structure of vanadium phosphate glasses was infrared absorption experiments.[715,722,745-748] The most complete study is the one of Anderson and Compton,[748] who studied V_2O_5 - P_2O_5 glasses containing 70% and 88% V_2O_5. In the former sample, vibrational absorption peaks were observed near 360, 420, 680, 900, 1010, and 1100 cm^{-1}; in the latter glass, the peaks occurred near 330, 435, 635, 810, 1010, and 1085 cm^{-1} Except for the peaks near 1010 cm^{-1}, the absorption bands have considerable width and overlap each other. The position of the peaks are independent of temperature in the 77 to 300K region.

Crystalline V_2O_5 exhibits vibrational absorption peaks near 915, 1040, 1260, and 1275 cm^{-1},[748-755] the peak at 1040 cm^{-1} being extremely sharp. The vanadium-oxygen stretching frequency has been reported to be in the 1025-1050 cm^{-1} range for compounds in which the vanadium atoms are totally ionized to V^{5+},[756,757] while the presence of V^{4+} ions appears to reduce the stretching frequency to the 900-1020 cm^{-1} range.[758,759] Anderson and Compton[748] concluded on this basis that the vibrational band near 1010 cm^{-1} in the glasses is due to V-O stretching vibrations. Also, since crystalline phosphate compounds exhibit strong vibrational peaks between 1100 cm^{-1} and 1400 cm^{-1},[760-764] and since these peaks appear to be shifted into the 1020 to 1100 cm^{-1} region in phosphate glasses,[765-768] Anderson and Compton[748] associated the 1085 to 1100 cm^{-1} peak in the vanadium phosphate glasses to the P-O stretching frequency.

Landsberger and Bray[769] studied the NMR and EPR spectra of several vanadium phosphate glasses, and were able to reach several interesting conclusions as to the basic structure of the material. They managed to locate and separate the V^{5+} and V^{4+} nuclear magnetic resonance lines, despite the fact that they are virtually superimposed, by means of a detailed analysis of the theoretical and observed line shapes. A calculation of the most probable value of the nuclear quadrupole coupling constant gave values in the 1.0 to 1.5 MHz range, close to the value in crystalline V_2O_5 of 0.8 MHz,[770] but quite far from the 2.9 to 4.4 MHz range observed in metavanadate compounds, in which the vanadium atoms are fourfold coordinated.[771] This is strong evidence that the vanadium atoms in the glass are fivefold coordinated, as in crystalline V_2O_5, in agreement with the conclusions from the infrared spectra. The question which then remains is how the PO_4 tetrahedra fit into the V_2O_5 structure. Landsberger and Bray[769] concluded from a detailed analysis of the composition dependence of the NMR chemical shifts that the most likely structure is one in which some of the VO_5 pyramids contain a PO_4 tetrahedron in place of its apex oxygen ion. If this is the case, charge neutrality suggests that two V^{4+} ions are formed for each PO_4 unit bonded in this manner. Also, since the V^{4+} NMR line exhibits an observable anisotropic hyperfine interaction, Landsberger and Bray concluded that the staying time of an excess electron on a vanadium ion was greater than the lifetime of the nuclear state, evidence for the partial localization of electrons of V^{4+} ions in the glass.

Landsberger and Bray[769] also investigated the EPR spectra of their glasses in order to determine the ratio of V^{4+} ions to V^{5+} ions as a function of composition, and they found that about one third of the vanadium atoms were quadruply ionized in $(V_2O_5)_{53} \cdot (P_2O_5)_{47}$. The ratio of V^{4+} to V^{5+} decreases monotonically to approximately 0.06 in $(V_2O_5)_{91} \cdot (P_2O_5)_9$, as might be expected. The trend of the ratio with composition parallels the estimates from chemical analysis.[772-775] However, they found that $(V_2O_5)_{43} \cdot (P_2O_5)_{57}$ showed a smaller V^{4+}/V^{5+} ratio than did the $(V_2O_5)_{53} \cdot (P_2O_5)_{47}$ glass, a surprising result that could be spurious.

Much remains to be sorted out with respect to the structure of vanadium phosphate glasses. Most important, it would be useful to ascertain how the VO_5 units link together. It can be assumed that a layer structure, as in crystalline V_2O_5, is predominant in V_2O_5-rich glasses, but a chain configuration, analogous to that shown for amorphous As_2Se_3 in Figure 22 (b), is another possibility. In glasses with high P_2O_5 concentration, the situation is clearly much more complex, in view of the three different crystalline forms of P_2O_5.

2. Electrical Properties

Because of the wide variations in composition

and stoichiometry, it is difficult to compare electrical data taken on different samples prepared in separate laboratories. Consequently, spectacular consistency has not been obtained in the several measurements of electrical conductivity as a function of temperature that have been reported on vanadium phosphate glasses.[746,747,773,776-779] Janakirama-Rao[746] studied complex glasses in the V_2O_5-P_2O_5-GeO_2 system, and found ordinary semiconductor behavior from 220K through 500K, with activation energies in the 0.2 to 0.4 eV range, decreasing with increasing V_2O_5 content. He obtained values for the pre-exponential factor, σ_o, between 10^{-1} Ω^{-1} cm^{-1} and 1 Ω^{-1} cm^1. Schmid[776] found similar behavior in the V_2O_5-P_2O_5 system in the 200 to 300K range, with $\sigma_o \sim 10^{-1} \Omega^{-1}$ cm^{-1} and activation energies varying from 0.24 eV to 0.44 eV, but he obtained breaks in the conductivity curves below 150K and evidence for conductivity increases relative to ordinary activated behavior near room temperature. In the 77 to 150K range, the log σ vs. T^{-1} curves flattened out in a somewhat irregular manner, and they do not improve their looks much on a $T^{-1/4}$ plot. The recent results of Linsley et al.[777] are shown in Figure 40. In these V_2O_5-P_2O_5 glasses, a reducing agent was used to control the V^{4+}/V^{5+} ratio at constant composition. It is clear that no real activation energy can be defined, although the apparent activation energy near room temperature is always in the 0.3 to 0.4 eV range. These curves do become considerably more linear on a $T^{-1/4}$ plot, but no conclusion can be drawn since the minimum temperature investigated was 50K, and some curvature can be noted even by that point. For these data, the apparent value of σ_o is of the order of 10 Ω^{-1} cm^{-1}, higher than in previous work, at least partially due to the increase in apparent activation energy above room temperature. In all the data reported, conductivity decreases with decreasing V_2O_5 content, despite the fact that the V^{4+}/V^{5+} ratio increases. Linsley et al.[777] found that at any constant composition, conductivity increased initially with increasing V^{4+}/V^{5+} ratio, but then exhibited a maximum near the point at which the V^{4+} ions represent between 10% and 20% of the total vanadium ions, depending on the composition of the glass. These maxima could be correlated with minima in the apparent activation energies, rather than with pre-exponential effects.

Thermoelectric-power measurements on vanadium phosphate glasses have been reported,[746,775,777,780,781] but agreement is not very good. At low values of the V^{4+}/V^{5+} ratio, the predominant conduction is always n-type. A change in sign of the thermoelectric power occurs for V^{4+}/V^{5+} ratios near unity, although Linsley et al.[777] found that the value of this ratio at the n-p transition point decreased with increasing V_2O_5 content. For any given glass, the thermoelectric power is essentially independent of temperature down to 200K;[775,777] at low temperatures, however, the thermoelectric power increases exponentially with decreasing temperature.[773] The magnitude of the thermoelectric power at a given V^{4+}/V^{5+} ratio varies considerably in the several measurements reported, a fact that cautions against a quantitative interpretation of the results.

The Hall mobility has been estimated to be of the order of 2 x 10^{-4} cm^2/V-sec at room temperature for a V_2O_5-P_2O_5-GeO_2 glass very rich in V_2O_5 (94%).[746] Neither the sign nor the temperature dependence were reported.

The ac conductivity of V_2O_5-P_2O_5 glasses is independent of frequency at high temperatures and low frequencies, and is essentially independent of temperature at high frequencies ($>10^8$ Hz).[777,782] At low temperatures, a linear increase in ac conductivity with frequency was observed in the 10^3 to 10^5 Hz region, but above 10^6 Hz an ω^2 region appeared, with signs of saturation occurring near 10^8 Hz. Recent work on amorphous As_2Se_3, discussed in section V, suggests that the ω^2 behavior is a contact effect rather than an intrinsic loss. The ac conductivity exhibits a maximum at the same V^{4+}/V^{5+} ratio as the dc conductivity, up to at least 5 x 10^7 Hz.

All interpretations of the electrical conduction process in vanadium phosphate glasses to date[727,776-780] involve the transport of localized electrons between the constituent vanadium ions. In these models, V^{4+} is considered to consist of a V^{5+} "core" plus an excess electron that is easily transferred to any essentially equivalent V^{5+} ion in the material, with or without the necessity of phonon assistance. It seems clear that if the conduction electrons are indeed localized, with a staying time longer than the average phonon frequency, then since the material is primarily ionic, a concomitant distortion of the ions in the vicinity of the V^{4+} ion must occur. In other words, small polarons, which were briefly discussed in section II, must form. Small-polaron transport, at least at high temperatures, can take place only with the assistance of phonons. Thus the mobility should exhibit thermal activation. In an ordered crystal, below about half the Debye temperature, it becomes more probable for small polarons to move via a tunneling process, by using the zero-point motion of the ions in what might be thought of as a polaron band.[783] In the band regime the mobility decreases with increasing temperature,[784] because of the deviations from perfect periodicity induced by the lattice vibrations. In a disordered glass, the polaron-band regime should be replaced by a phonon-assisted tunneling regime, in which a $T^{-1/4}$ law behavior of conductivity, c.f. Equation 28, predominates.[785] There is some evidence that the conductivity of vanadium phosphate glasses does exhibit this type of behavior, as can be seen from Figure 40. However, just such a temperature variation of conductivity is observed, for example, in crystalline NiO,[786] and merely represents a transformation from impurity conduction at low temperatures to ordinary band-like conduction at high temperatures. Thus, no firm conclusions should be drawn from this behavior alone.

The dependence of the conductivity on the V^{4+}/V^{5+} ratio for a given composition can also be used to test the small-polaron model. If c is the ratio of the number of V^{4+} ions to the total number of vanadium ions in the glass, then the conductivity at high temperatures should be[787]

$$\sigma = \frac{A}{(kT)^{3/2}} c(1 - c) e^{-E_p/2kT} \qquad (41)$$

where A is a constant and E_p is the binding energy of the polaron. Equation 41 has a maximum at c = 0.5, the point at which the concentrations of V^{4+} and V^{5+} are equal. This is not consistent with the data,[777] in which the maximum in conductivity occurred for values of c between 0.1 and 0.2. Linsley et al.[777] suggested that not all the V^{5+}

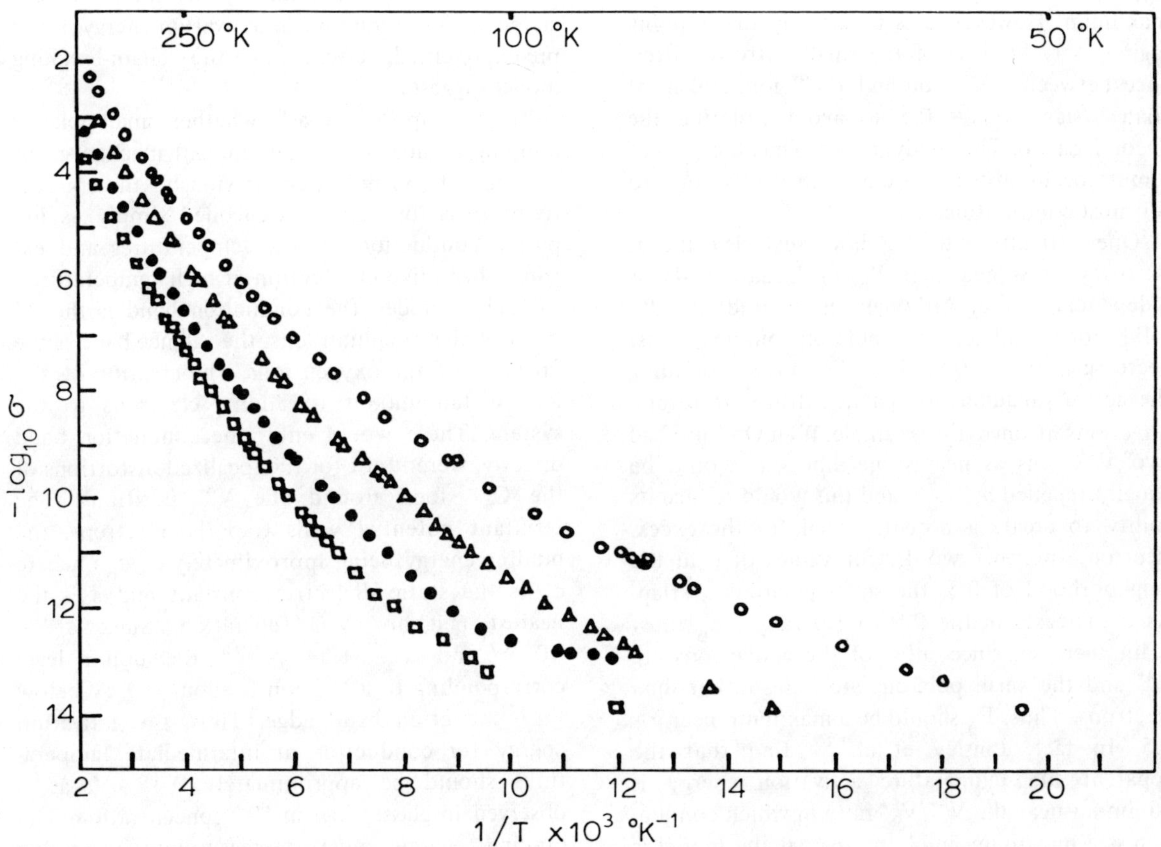

FIGURE 40. Conductivity as a function of temperature for vanadium phosphate glasses: (○) 90% $V2 0_5$; (Δ) 80% $V_2 O_5$; (●) 70% $V_2 O_5$; (□) 60% $V_2 O_5$; (Linsley et al.[777]).

ions in vanadium phosphate glasses are equivalent, due to the tendency of V^{5+} to form polymeric-type ions.[788] If a large percentage of the V^{5+} ions were present in the glasses as, for example, trimeta-phosphate-vanadate rings, then these would be inert as far as the conductivity is concerned. In reality, the parameter c in Equation 41 should be the ratio of V^{4+} ions to total vanadium ions in equivalent, i.e. VO_5, environments.

In the small-polaron hopping regime, the thermoelectric power should take the form,[789]

$$S = \frac{k}{e}\left[\ln \frac{c}{1 - c} + a\right] \quad , \qquad (42)$$

where a is a small constant, essentially zero.[790-793] Equation 42 predicts that the thermoelectric power changes sign at c = 0.5, which is consistent with the data only if all V^{5+} ions are equivalent. Thus, the interpretations of the conductivity and thermoelectric results are inconsistent. Linsley et al.[777] note that a large negative value of a in Equation 42 could allow the thermoelectric-power data to change sign for c in the 0.1 to 0.2 range, where the conductivity is maximum. However, a is nonzero in the hopping regime only because of the small entropy difference between a V^{5+} ion and a V^{4+} ion, and must change sign at c = 0.5 to avoid violating the Second Law of Thermodynamics. Thus the zero of s must occur just at c = 0.5, even if a is nonzero for most compositions.

One difficulty with the above analysis is that it is always assumed that E_p in Equation 41 is independent of c. Although this is undoubtedly valid for small c, the polaron binding must decrease near c = 0.5, since the ions surrounding the active vanadium ion cannot distort in several directions at once. For example, if an O^{2-} ion had two V^{4+} ions as nearest neighbors it would be equally repelled by both, and this would reduce its ability to create a potential well for the excess electron. In other words, for values of c in the neighborhood of 0.5, the small polarons overlap. As c increases in the 0.5 to 1.0 range, E_p must again increase, since most of the active ions are V^{4+} and the small polarons are holes rather than electrons. Thus, E_p should be a maximum near c = 0.5. In fact, Linsley et al.[777] find that the apparent high-temperature activation energy is minimum near the V^{4+}/V^{5+} ratio in which conduction is a maximum, implying that, if the model is appropriate, this is the region that should be associated with c = 0.5. It is thus the thermoelectric data rather than the conductivity data that appear to be inconsistent with the small-polaron model.

Linsley et al.[777,795] also proposed an alternative model, suggested by the work of Ohashi,[794] that polymeric low-resistance chains form in the glass, but that V^{4+} ions break these chains and introduce high series resistance between the chains. If the hopping model remains valid in such a structure, the conductivity will exhibit a maximum at much smaller values of c than 0.5, since the additional high resistance introduced by each V^{4+} ion created would very shortly more than compensate for the additional carrier. Since the high resistances limit the mobility rather than the concentration of carriers, the thermoelectric power should still change sign at c = 0.5, in agreement with experiment. However, this model has difficulties in accounting for the sharpness of the conductivity maximum and the lack of significant frequency or temperature variations of the concentration at which the maximum occurs. Furthermore, the conductivity maximum is associated with a minimum in activation energy, not a pre-exponential effect, as the chain-breaking model suggests.

It is tempting to ask whether small-polaron hopping is indeed the predominant mechanism of conduction. As indicated previously, the conductivity behavior can be explained simply as impurity conduction at low temperatures and extrinsic band-like conduction at high temperatures. In such a model, the conduction band is the 3d band of the vanadium ions, the valence band is the 2p band of the oxygen ions. The creation of V^{4+} ions is tantamount to adding electrons to the system. These would enter the conduction band directly, were it not for the localized distortions of the O^{2-} ions around the V^{4+} positions. The resultant potential wells trap the electrons, the binding energy being approximately $e^2/\epsilon_o a$, where ϵ_o is the static dielectric constant and a is the nearest neighbor V-O separation. Since $\epsilon_o \approx$ 10,[777] and a = 1.54 Å,[796] the donor level corresponding to a V^{4+} ion is about 0.9 eV below the conduction band edge. Thus, the activation energy for conduction at intermediate temperatures should be approximately 0.45 eV, as is observed in glasses low in V^{4+} concentration. The binding energy must decrease with increasing density of V^{4+} ions, since the oxygen ions cannot distort around two V^{4+} ions simultaneously, as discussed previously. This is also in agreement with observation. The material can be assumed to be partly compensated via either impurities, structural defects, or even vanadium vacancies. Thus, impurity conduction dominates at sufficiently low temperatures. As the V^{4+}/V^{5+} ratio increases for a given composition, the extent of the compensation increases, resulting in a decrease of the Fermi energy relative to the conduction band edge at a given temperature. This leads to a minimum in the room-temperature activation energy, which in turn results in a conductivity maximum. When the V^{4+}/V^{5+} ratio exceeds unity a different approach is necessary, since the short-range order is probably more like that of VO_2 than V_2O_5. In this case the V^{5+} ions act like V^{4+} ions with an excess hole. The O^{2-} ions then distort and trap the excess holes, and these traps become acceptors for the lowest 3d band, the valence band of VO_2.[797] Conduction is then p-type, in agreement with the thermoelectric power data.

For the above model to be applicable, the 3d bands of V_2O_5 and VO_2 must exhibit a mobility edge and possess delocalized states. A guide to the nature of the conduction band in vanadium phosphate glasses can be obtained from a study of crystalline V_2O_5. Early work on crystalline V_2O_5 was interpreted in terms of a small-polaron hopping model,[798] but it now appears clear that band-like conduction in the 3d band predominates[799,300] and that the donor-like V^{4+} ions arise primarily from impurities rather than oxygen vacancies.[801-804] Crystalline VO_2 is a semiconductor at low temperatures, but it undergoes a semiconductor-metal transition at 340K.[797] The nature of the low-temperature state has not yet been unambiguously determined, and suggestions have been made that it represents an ordinary semiconductor[805] or a Mott insulator.[806] In any event, the 3d bands of both V_2O_5 and VO_2 are not far removed from the insulating side of the Mott transition, and the additional disorder resulting from the lack of long-range periodicity in the corresponding amorphous forms could well lead to a complete localization of the states in the band.[123] Amorphous V_2O_5 films have been prepared by rf sputtering and have a room-temperature resistivity about a factor of 10^2 larger than crystalline V_2O_5 films.[711] However, no conclusion can be drawn as yet about the conduction mechanism. Sputtered VO_2 films are either metallic or insulating, depending on whether the substrate temperature is greater or less than 340K.[807] In either case, the semiconductor-metal transition does not occur in the amorphous films.[808] One possible conclusion is that, for the short-range order corresponding to the low temperature structure, the positional disorder induces a complete localization of the 3d bands. (On the other hand, the fact that metallic conduction can occur in films deposited at a high substrate temperature, despite the lack of periodicity, makes this somewhat unlikely.) In any event the addition of compositional disorder introduced by the P_2O_5 is a further factor that could lead to complete localization in the 3d bands of the vanadium phosphate glasses. If so, conduction at all temperatures can proceed only with the assistance of phonons. A generalization of the previous model is still possible provided that the V^{4+} ions in the glass exist as vanadyl (VO^{2+}) ions. Then the activation energy for conduction would largely represent the energy necessary to free an electron from a vanadyl ion and place it on a neighboring V^{5+} ion. Once there, it forms a small polaron and can contribute to conduction by means of hopping between equivalent V^{5+} sites. The small-polaron hopping energy then also contributes to the activation energy, although there is a possibility it is quite small.[52] There is some experimental evidence in favor of such a model. The ac conductivity at room temperature in the 1 to 100 MHz range increases approximately proportional to frequency, a behavior that has been ascribed to a hopping component of conduction. But in this frequency range, the ac conductivity exhibits a maximum at the same V^{4+}/V^{5+} ratio as the dc conductivity, an indication that the two processes are strongly related.

3. Optical Properties

Optical-absorption experiments on vanadium phosphate glasses in the visible have been carried out by Anderson and Compton,[748] and a typical result is shown in Figure 41. The optical gap is about 2.6 eV, and the slope of the absorption edge is approximately 0.06 eV, a typical Urbach's-law value. The value for the optical gap is slightly larger than that for crystalline V_2O_5, which exhibits a dichroism due to its anisotropy, an Urbach edge, and most likely a direct forbidden gap.[749-753] Anderson and Compton found that absorption near the edge in the vanadium phos-

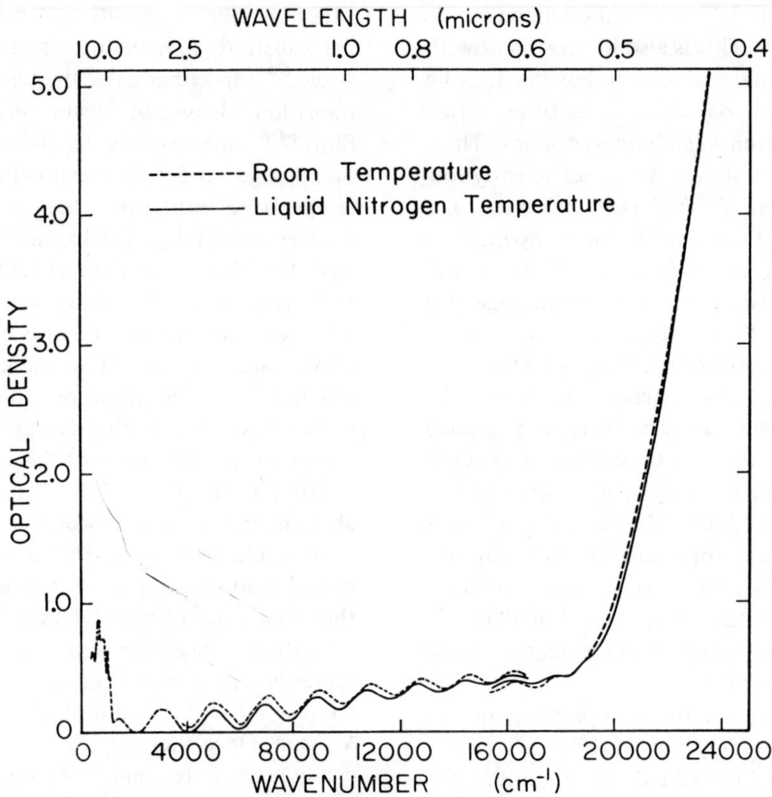

FIGURE 41. Optical absorption spectrum of a vanadium phosphate glass with 87.5% V_2O_5 at 77K and 300K (Anderson and Compton[748]).

phate glasses also fits the relation for direct forbidden transitions,[809]

$$\alpha \propto (\hbar\omega - E_g)^{3/2}$$

In fact, the similarity between the absorption edges in the glasses and in crystalline V_2O_5 is striking.

Anderson and Compton[748] also found a series of absorption peaks between 0.2 eV and 2.5 eV, as can be seen in Figure 41. These are not observed in crystalline V_2O_5.[749-755] Although it is difficult to decide with certainty as to their origin, they could well arise from the crystalline-field peaks[810] of the V^{4+} ions present in the glass. The O^{2-} ions in the glass form a crystalline field of low symmetry, and the V^{4+} concentration is such that the value of the absorption constant at the peaks is consistent with that expected from the forbidden d-d transitions. However, about 10 distinct peaks are observable, the theoretical upper limit for a tenfold degenerate 3d′ ion. But since the Kramer's degeneracy[811] cannot be split by either the crystalline field or spin-orbit coupling, at most five independent peaks can be due to excitations of V^{4+} ions, if they all are in equivalent positions. On the other hand, any other simple electronic transitions, such as those from the 3d′ state to the 4s states of the vanadium ions or charge-transfer excitations should be considerably higher in energy.[122] It is thus likely that the V^{4+} ions have at least two distinct but different environments in the vanadium phosphate glasses, perhaps as vanadyl ions and in more complex configurations involving phosphate subunits.

4. Conclusions

It is clear that little can be concluded with certainty at this time. Until a less opaque view of the structure of the glasses is obtained, most of the models for the electrical and optical properties must remain speculative. As opposed to the situation in the elemental and covalent amorphous semiconductors that have been discussed previously, the activation energy for electrical conduction in vanadium phosphate glasses is not approximately half the optical gap, but is considerably smaller. This is because the optical gap is sufficiently large, 2.6 eV, that intrinsic semiconduction is insignificant below the crystallization temperature. It is clear that the extrinsic conduction which predominates at room temperature is due to the presence of V^{4+} ions in the glasses, but it is not all obvious yet whether the conduction mechanism is by means of transport in an ordinary band or via hopping of small polarons. Several speculative models have been suggested to account for the available transport data.

B. Iron Phosphate Glasses

The only primarily ionic glasses other than those based on the V_2O_5-P_2O_5 system whose electrical properties have been studied to any extent are those based on the FeO – P_2O_5 system. Unfortunately, very little information is known about the structure of these glasses. However, it is painfully evident from a consideration of the crystalline phases of the main constituents of the glasses that the structure must be exceedingly complex. As discussed in subsection A, P_2O_5 crystallizes in three different forms, although all are based on PO_4 tetrahedra. The situation with regard to iron oxide is still more complicated. Iron commonly exists in either of two valence states, Fe^{2+} ($3d^6$) and Fe^{3+} ($3d^5$). Three different iron oxides are known, FeO, Fe_3O_4, and Fe_2O_3. FeO is always non-stoichiometric, with at least 2% excess oxygen.[812] Above 203K, the structure is that of NaCl; a magnetorestrictive-type distortion occurs below the magnetic ordering temperature.[813] Fe_3O_4 has the spinel structure above 119K;[314] below that temperature, a small distortion to orthorhombic symmetry occurs.[815] The material contains Fe^{3+} and Fe^{2+} ions in a 2:1 ratio; half of the Fe^{3+} ions have tetrahedral environments, half have octahedral environments.[816] Fe_2O_3 has three crystalline forms, with rhombohedral,[817] cubic,[818] and tetragonal[819] symmetry. The short-range order is different in all three structures. It is difficult even to speculate on the structure of the iron phosphate glasses at this time, but it is very likely that there is a multiplicity of environments for both the Fe^{2+} and the Fe^{3+} ions.

Electrical conductivity has been measured in $(FeO)_{55} \cdot (P_2O)_{45}$ for several Fe^{3+}/Fe^{2+} ratios,[820,821] with some typical results shown in Figure 42. The conductivity exhibits thermally activated behavior from room temperature through 700K, with activation energies in the 0.6 to 0.7 eV range. There is evidence for a break in the log σ vs. T^{-1} curves near 400K, producing a small increase (approximately 0.1eV) in activation energy at higher temperatures. The value of the pre-exponential factor, σ_0, extrapolated from the high-temperature behavior is about 10 Ω^{-1} cm^{-1}, essentially independent of the Fe^{2+}/Fe^{3+} ratio.

A maximum in conductivity occurs near an Fe^{3+}/Fe^{2+} ratio of unity, just what would be expected from Equation 41 if all iron sites were equivalent. This is distinctly different from the results obtained in vanadium phosphate glasses.

The thermoelectric power of iron phosphate glasses is independent of temperature from 400K through 600K, but is very strongly dependent on the Fe^{3+}/Fe^{2+} ratio.[820] For small Fe^{3+}/Fe^{2+} ratios, the sign of the thermoelectric power indicates the predominance of p-type conduction, but a transition to n-type conduction occurs near an Fe^{3+}/Fe^{2+} ratio of 0.5. This is not consistent with Equation 42 if all the iron sites are equivalent.

For the vanadium phosphate glasses, the thermoelectric-power sign change is where it might be expected, but the conductivity maximum occurs at too small a value of c. The opposite problem characterizes the iron phosphate glasses, in which the conductivity maximum occurs at c = 0.5, but the sign of the thermoelectric power changes at a significantly smaller value.

Kinser[821] annealed some samples of $(FeO)_{55}$ $(P_2O_5)_{45}$ at various temperatures for one hour, and observed small conductivity increases for annealing temperatures up to 800K. Annealing in the 800 to 1100K range induced large decreases in conductivity. Electron microscopy indicated the appearance of fine-grain crystals once the annealing temperature reached about 700K. The crystals appeared to be $Fe_3(PO_4)_2$. Dielectric-loss measurements showed loss peaks in the annealed samples concomitant with the appearance of fine-grained crystallites; these peaks disappeared when the crystal size became macroscopic. The loss peaks thus appear to be Maxwell-Wagner[822,823] losses, due to a more conductive dispersed phase, and should not be considered as evidence for a hopping component of conduction. If the crystallites are indeed $Fe_3(PO_4)_2$, then they tend to make many of the Fe^{2+} ions in the material inert as far as conduction is concerned, thus increasing the effective Fe^{3+}/Fe^{2+} ratio and in turn the electrical conductivity.

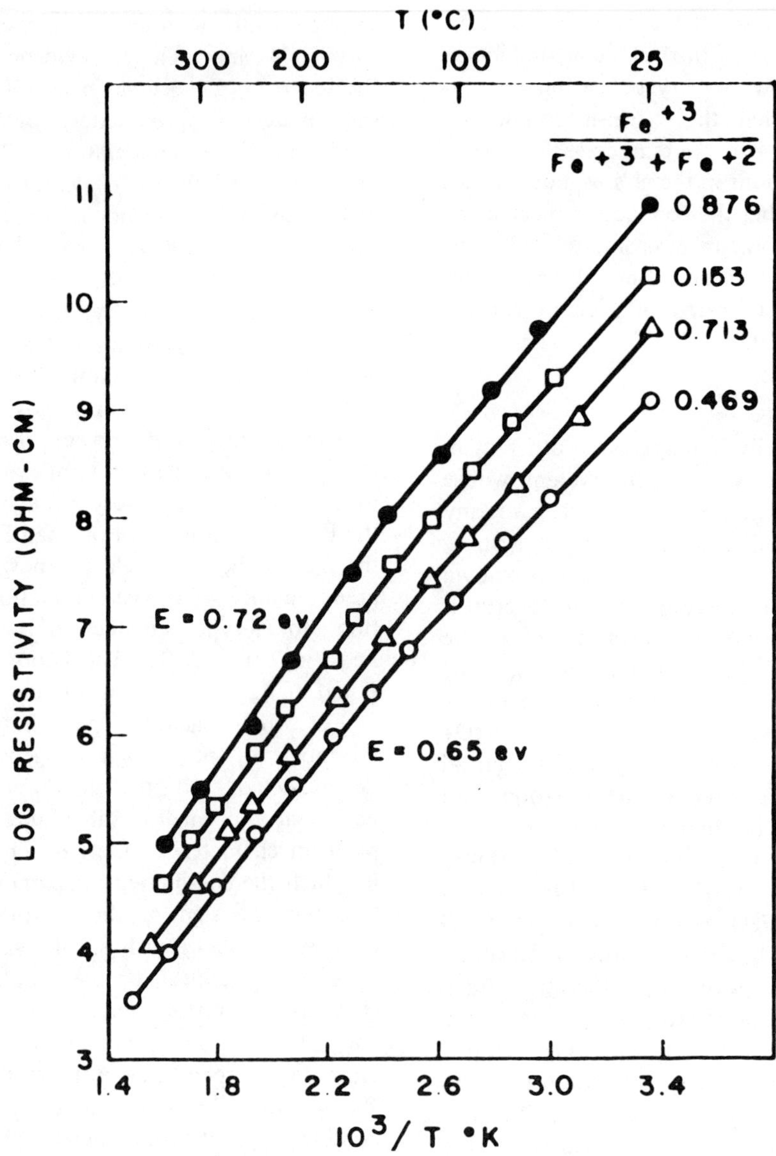

FIGURE 42. Resistivity as a function of temperature for amorphous $(FeO)_{55}\cdot(P_2O_5)_{45}$ with various Fe^{3+}/Fe^{2+} ratios (Hansen[820]).

VII. HIGH-FIELD EFFECTS AND SWITCHING PHENOMENA

Application of sufficiently high electric fields to any material sooner or later leads to deviations from linearity in the resultant current. Such deviations are referred to as non-ohmic conductivity. There are two classes of explanations for such effects – thermal and electronic. The former arise because of the fact that interactions between electrons and phonons exist in all materials. When electrons are accelerated by the field, they can always emit phonons, thus giving up some of their excess energy to the ion cores. This increases the temperature of the material, an effect known as Joule heating. Electronic non-ohmic effects arise because of changes in the response of the electrons in the material to the electric field itself at a constant temperature. Examples of electronic effects are field-induced tunneling across an energy gap or from a localized trap into the conduction band (Zener tunneling),[824] thermal excitation of an electron from a trap over a field-reduced Coulomb barrier (Poole-Frenkel effect),[389] tunneling from a metallic electrode into the conduction band (Fowler-Nordheim emission),[825] or thermal excitation from an electrode over a field-reduced Schottky barrier (Schottky emission).[826] Electronic effects become possible in the more insulating solids only because the Joule heating at a given electric field, being proportional to the conductivity of the material at a constant field, is usually negligibly small. These high-field effects are generally investigated in thin films rather than in bulk materials for the simple reason that even modest voltages, such as 100V, can be used to induce high fields, of the order of 10^6 V/cm, across a 1μ film. The low-field conductivity of insulating materials is often so small that high-field experiments provide the primary means of investigating the electrical properties. Several excellent review articles which deal with non-ohmic conduction in dielectric films have appeared recently,[827–833] and we shall not deal with these effects in any detail in this review. However, when analyzing high-field behavior with regard to switching phenomena, we must pay careful attention to such items as whether the contacts are injecting, neutral, or blocking, and whether the current is bulk-limited or electrode-limited.

Although non-ohmic effects are scientifically interesting and have been used to reach vital conclusions about the bulk density of states and carrier mobilities in several amorphous semiconductors, e.g. Se and As_2S_3, it is the negative-resistance and switching effects that are of primary importance in the materials under discussion. These effects can be divided into six classes,[834] as shown in Figure 43. Figure 43 (a) shown the I-V characteristic of a voltage-controlled negative resistance (VCNR) device, also called an N-shaped instability. Figure 43 (b) is the characteristic of a VCNR device with memory, in which both a low-resistance and a high-resistance state exists in the absence of an applied field. The I-V characteristic of a current-controlled negative-resistance (CCNR) device, or S-shaped instability, is shown in Figure 43 (c). A switching device, shown in Figure 43 (e), exhibits an I-V characteristic very similar to

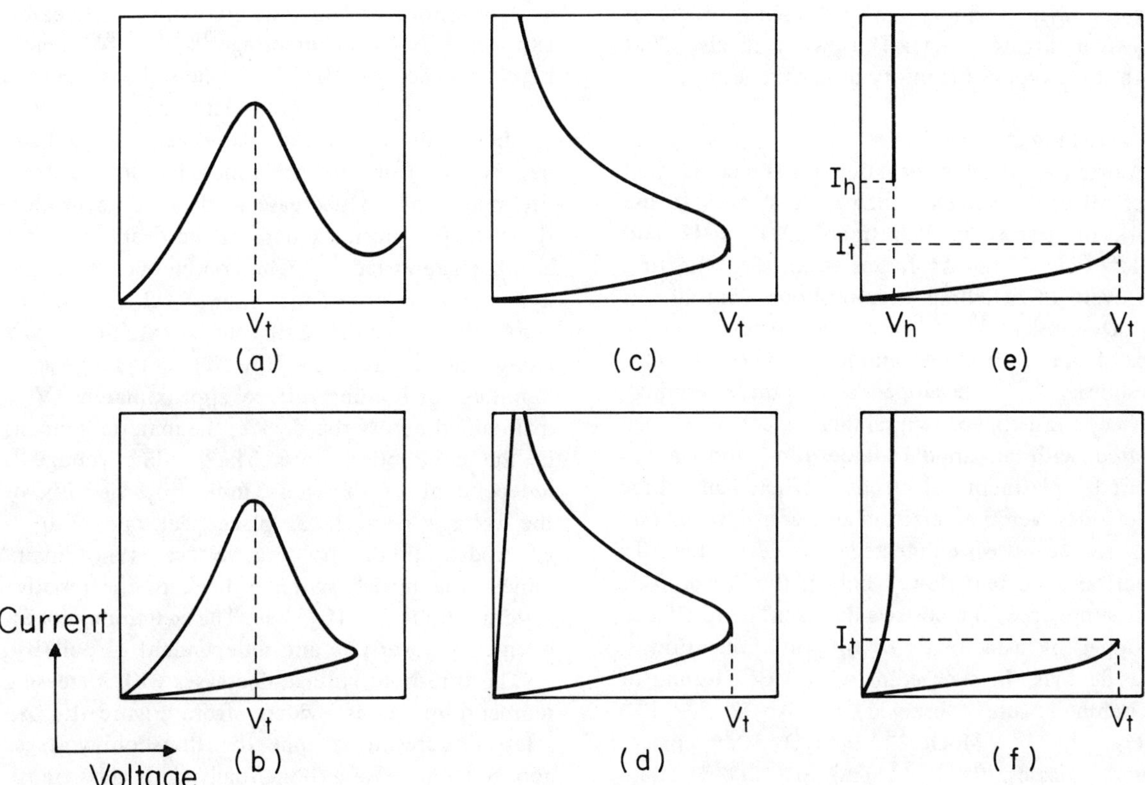

FIGURE 43. Six possible types of negative-resistance behavior observed in amorphous solids (a) VCNR; (b) VCNR with memory; (c) CCNR, (d) CCNR with memory; (e) threshold switching; (f) memory switching.

that of a CCNR device, except that no stable operating state exists along the negative-resistance region. A CCNR device will ordinarily switch along the load line from the high-resistance, or OFF, state to the high-current, or ON, state, in which case its characteristic will be indistinguishable from that of a switching device. In principle, a sufficiently large load resistance should allow the stabilization of a CCNR device at any point on the negative-resistance branch, whereas a switching device would only oscillate under such operating conditions. However, as Shaw and Gastman[835] have recently shown, the device capacitance and inductance are sufficient to induce oscillations in a CCNR device, and thus it is essentially impossible to differentiate CCNR from switching devices. Consequently, we shall treat them together, bearing in mind that fundamental differences between the two could manifest themselves in respects other than the I-V behavior. Figure 43 (d) shows the I-V characteristic of a CCNR device with memory, in which two distinct states exist, a high-resistance and a low-resistance state, even in the absence of an applied field. In practice, these devices are difficult to distinguish from switching devices with memory, whose characteristic is shown in Figure 43 (f). Thus we shall also treat both these types of memory devices together.

A. Switching Devices

Current-controlled negative resistance is well over 50 years old, stemming at least back to the work on boron by Weintraub[836] in 1913 and Lyle[837] in 1918. As far as is known, the first discovery of switching in amorphous material was by Ovshinsky,[838,839] who constructed a device from a 900 Å anodic film of Ta_2O_5. Ovshinsky[840] developed a three-terminal device, consisting of two tantalum electrodes, each coated with an anodic oxide film, and an uncoated, platinum electrode. When all three electrodes were immersed in an electrolyte, and an ac source was connected between the Ta electrodes, current flowed only if the Pt electrode was supplied with a positive dc bias. The resistance dropped by a factor of 2000 upon application of the dc bias. In succeeding years, CCNR and/or switching were observed in Nb_2O_5,[841-844] SiO_2,[845-847] MoO_3,[848] Al_2O_3,[849] chalcogenide glasses,[1,850-877] and Si.[878,879] There has also been a wealth of reports of switching in crystalline solids,[880-891] in organic films,[892] in liquids,[893] in heterojunctions,[894] and in p^+ - n - n^+ junctions.[895] Obviously, there are enormous variations in the characteristics, parameters, and quality of switching in all of these materials, and many different fundamental mechanisms can be involved. We shall not attempt to review the details of switching in all of these materials, but rather shall concentrate on the most intensively investigated class, the chalcogenide glasses, particularly those in the Te-As-Si-Ge system.

1. Experimental Results

A typical chalcogenide-glass switching device is 1-10μ thick, sandwiched between two electrodes.[1] When low voltages are applied, conduction is ohmic, with resistances of the order of $10^5\,\Omega$. Above fields of about 10^4 V/cm, non-ohmic processes become evident, and the current rises exponentially with applied voltage.[861] This non-ohmic conduction appears to be bulk-limited.[1] Switching occurs at a critical field, generally of the order of 10^5 V/cm. The actual switching is extremely rapid, occurring in less than 1.5 x 10^{-10} sec.[1] However, there is a delay time prior to switching, typically about 10 μsec at threshold. The delay time decreases exponentially with overvoltage,[834,862,896] reaching the order of 10^{-9} sec when the voltage is about 50% above threshold. This type of behavior is shown in Figure 44. However, the voltage dependence of delay time is composition-dependent, and a decrease in delay time proportional to the inverse square of the overvoltage has also been reported.[858] The conductivity state has dynamic resistance of the order of $1\,\Omega$. The voltage across the device drops sharply as switching occurs along the load line. However, as long as a minimum or holding voltage, approximately 1V, is maintained across the device, the material remains in the conducting state. The holding voltage is independent of thickness, indicating the bulk of the voltage drop takes place near one or both electrodes. If the holding voltage is not maintained, the device switches back to the resistive state in about 5 x 10^{-7} sec. The switching process is entirely reversible and independent of polarity.

The threshold voltage decreases with increasing temperature, as is evident from Figure 45. An interesting result is that the threshold voltages appear to increase exponentially with glass transition temperature for a large class of materials.[874] Threshold voltage for a given device decreases with

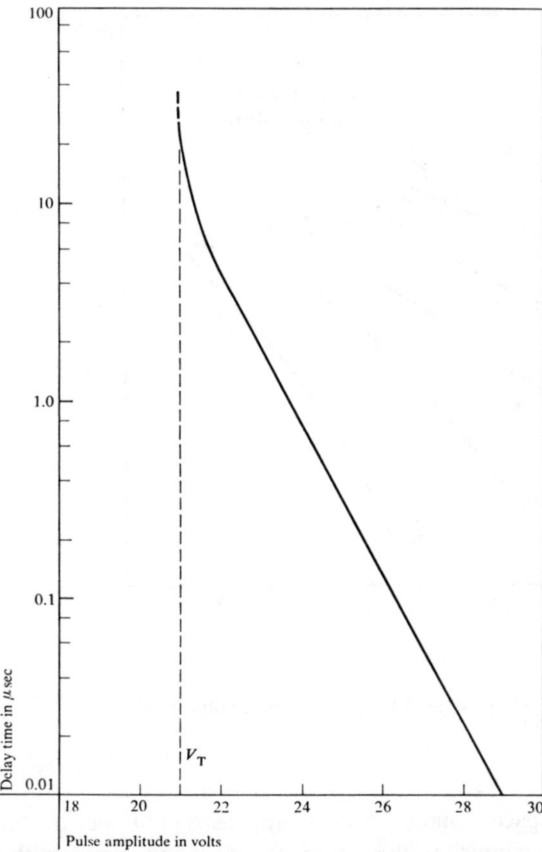

FIGURE 44. Delay time between the onset of a square-wave pulse and switching as a function of pulse height (Fritzsche[834]).

increasing frequency,[869] and exponentially decreases with increasing pressure.[879,898]

It appears that a forming process is necessary prior to switching in most if not all devices.[833,899] Prior to formation, the current is non-ohmic, but no negative-resistance appears until a critical current density is exceeded. This occurs at a forming voltage, V_F, which is generally about 30% larger than the threshold voltage of the formed device. Once formed, the threshold voltage is stable. Little is known about the forming process, but it most likely represents an effect connected with the electrode-semiconductor interfaces.

Negative capacitance just prior to threshold has been observed in chalcogenide glasses.[900] The negative capacitance is not an inductance effect, since it is independent of frequency from 15 KHz to 100 KHz. Below 240K, the capacitance appears to be negative even with no applied voltage. On the other hand, Shaw[901] observed only positive capacitance, 1.5 to 2.5 pF, independent of bias, from 200K to 350K and from 15 KHz to 5 MHz for a large number of chalcogenide-glass switches. Rockstad[902] has pointed out that intrinsic generation of electrons and holes in equal concentrations in low-mobility materials cause the current to lag behind the voltage, thus giving a negative contribution to capacitance. The negative capacitance arises because of the finite time necessary to redistribute the steady-state space charge after excess carrier generation, and is independent of frequency provided ω^{-1} is large compared to the electron and hole transit times. This model cannot account for the zero-bias negative capacitance at low temperatures. However, because of the results of Shaw, it is clear that until the negative-capacitance effects are experimentally reproduced, any interpretation is speculative.

There is much evidence that conduction in the ON state does not take place uniformly, but rather occurs along filamentary paths.[1,857,871] This is in accordance with a general result of Ridley,[903] who showed that CCNR always leads to high-current filaments, provided that the minimum-entropy-production principle of irreversible thermodynamics is applicable.

2. *Mechanisms*

In this subsection, we shall attempt to describe and categorize all mechanisms which can conceivably induce I-V characteristics such as those of Figure 43 (b) or Figure 43 (c), whether or not they have yet been proposed to apply to threshold-type switching in chalcogenide-glass films. In order to explain the conducting state, we must account for a sharp, non-equilibrium increase of the conductivity of the amorphous semiconductor. Since conductivity is essentially a product of a carrier concentration and a mobility, an immediate classification scheme for switching mechanisms presents itself – switching can be induced by (a) a sharp increase in carrier concentration, (b) a sharp increase in mobility, or (c) sharp increases in both.

Since switching is field-induced, we must first consider the effects of an applied electric field on the device, a chalcogenide-glass film sandwiched between two electrodes. To begin with, an electric field certainly accelerates any free carriers that have been thermally excited beyond the mobility edges. These carriers have reasonably high mobility, but they are present in extremely low

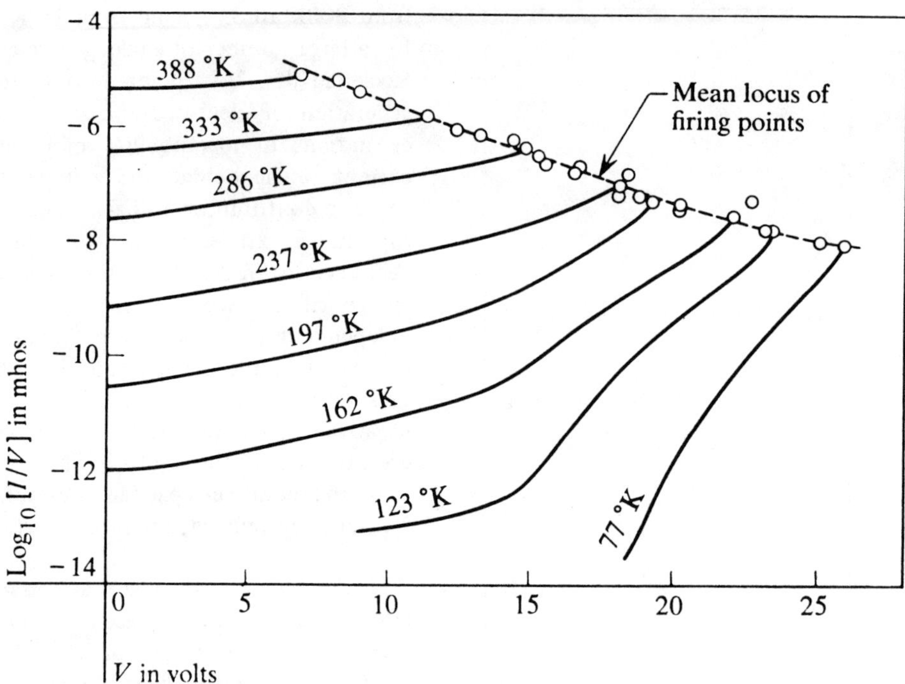

FIGURE. 45. Conductance of a 0.8 μ chalcogenide glass as a function of bias voltage. The conductance is independent of polarity (Fritzsche[8 34]).

concentrations at room temperature. As the carriers are accelerated they pick up energy from the field and thus move farther from the mobility edge. If a carrier can attain an energy relative to the mobility edge that is larger than the electrical activation energy, it can then excite another free carrier beyond the mobility edge and still maintain sufficient energy to remain free itself. This is the phenomenon of impact ionization, which is often a source of non-ohmic conduction in crystalline semiconductors. Fighting the occurrence of impact ionization are recombination mechanisms including non-equilibrium trapping and thermalization, i.e. the surrender of excess energy to the ion cores by emission of phonons, thus preventing the electronic temperature from rising above the actual temperature of the material. However, for a sufficiently large value of the applied field per unit mean free path, thermalization cannot occur sufficiently rapidly for the electrons to remain at the temperature of the ion cores, and the electrons become "hot." Eventually, a chain reaction of impact ionizations take place, the phenomenon of avalanche breakdown. If space-charge effects are unimportant, the field remains essentially homogeneous across the material, and at most a small negative resistance appears. However, if a large space charge builds up, as could occur for example, if the Fermi energy is much nearer the valence-band mobility edge than the conduction-band mobility edge of the hole mobility greatly exceeds the electron mobility, the field will become large near one of the electrodes. Since the critical field then need only be exceeded in the high-field region in order to maintain the large carrier concentration, the holding voltage can be much lower than the threshold voltage, and a large negative resistance can be obtained.

A second possibility for a sharp increase in carrier concentration is by means of tunneling, either from the electrodes, from filled traps, or across the mobility gap. Tunneling can occur even at zero temperature, so it does not require the existence of equilibrium carriers, as does impact ionization. If the CFO model is appropriate, tunneling from the Fermi energy to the mobility edges predominates. Since the tunneling rate increases exponentially with increasing field, a critical field exists at which the generation rate exceeds the recombination rate, thereby producing a sharp increase in carrier concentration known as Zener breakdown. Zener breakdown is not usually characterized by a negative-resistance region, since a reduction in field results in a failure to maintain breakdown conditions, and rapid recombinations return the device to the resistive state.

Still another means of sharply increasing the carrier concentration is by means of Joule heating, which, for a given applied field, is proportional to the conductivity of the material. Since the conductivity of chalcogenide glasses increases exponentially with temperature, the Joule heating will provide rapid rises in both temperature and current once the heat production becomes too intense to be dissipated to the electrodes by thermal conduction. This is known as thermal runaway. Once the critical current for thermal runaway is attained, the steady state is achieved only after a large temperature rise and thus a sharp decrease in device resistance. Under certain conditions, this can produce a negative-resistance region.

Finally, the tendency of an electric field to align dipoles can lead to a sharp increase in carrier concentration in a partially covalent, partially ionic glass. The covalent component of binding requires given bond angles in order to maintain the cohesive energy of the glass, and at low fields this dominates any energy decrease that could result from dipole reorientation. However, above a critical field, the energy balance reverses and a non-equilibrium ferroelectric-type phase transition is induced. Since this structural change breaks many of the covalent bonds, it is equivalent to excitation of free carriers, and leads to a sharp increase in conductivity.

There are several methods by which a sharp mobility increase can be induced by an applied field, but all of them depend on the mobility being low at equilibrium. Such would be the case if the amorphous semiconductor were, for example, a Mott insulator in which the predominant conduction mechanism was via small-polaron hopping. As discussed in section II, a Mott insulator exists because the kinetic-energy decrease available to an electron upon delocalization does not overcome the potential-energy increase that would result from the Coulomb repulsion between two electrons simultaneously present on the same ion core. But the excess free carrier generated by an applied field are available to partially screen out the Coulomb repulsion, and eventually reverse the energy balance, resulting in what might be called a non-equilibrium Mott transition. The field-generated carriers also partly screen the Coulomb interactions that lead to self-trapping of carriers and polaronic binding in ionic glasses, and thus an applied field could induce a small-polaron decoupling, sharply increasing carrier mobility. Finally, the fluctuating potential arising from long-range disorder is also screened, and this could lead to an effective decrease of the mobility gap with applied field. Such a model can be shown to result in a sharp increase in conductivity,[904] and can be called a non-equilibrium Anderson Transition.

Several models for amorphous-semiconductor switching depend on the applicability of the CFO model. This model has the advantage that a non-equilibrium conductive state can be envisioned in a simple manner. If a mechanism of double injection of electrons and holes can be supported in a situation in which all the charged traps in the mobility gap are filled, then the material will exhibit high conductivity. This is because the densities of positive and negative traps in the CFO model are essentially equal, and no space-charge region exists in the trap-filled limit. Thus, the carrier mobility will be relatively high, and there will be no Coulombic barrier for either electron or hole conduction.

3. Comparison of Theory with Experiment

The models enumerated in the previous subsection can, to an extent, be tested against the available experimental results. One pitfall of this procedure is that different compositions and geometries can lead to different switching mechanisms, so that use of given data to eliminate a model only eliminates it in one particular case. Nevertheless, the experimental observations are often quite general, provided we restrict ourselves to thin-film sandwich structures.

Avalanche breakdown does not, at first sight, appear to be a particularly promising mechanism for switching in chalcogenide glasses. The activation energy for conduction is about 0.5 eV, the mean free path of carriers is approximately 5 Å, and the threshold field is of the order of 10^5 V/cm. Thus, the carrier acquires an energy of only 0.005 eV or thereabouts between collisions. Consequently, a carrier must survive an average of 100 collisions without significant loss of energy in order for a single impact ionization to take place. It would thus appear that even avalanche multiplication, much less avalanche breakdown, is extremely unlikely in amorphous solids. However, Hindley[905] has suggested that hot electrons are possible in chalcogenide glasses, even at fields of

10^5 V/cm, since most of the collisions that limit the mean free path are elastic. Mott[905] has proposed that this comes about because phonons cannot be produced at a rate faster than the maximum phonon frequency, ω_{max}. Otherwise, they could not be carried away sufficiently rapidly to avoid piling up. Thus, the maximum rate of loss of energy to phonons is $\hbar\omega_{max}{}^2$. Furthermore, since the mobility and the mean free path very likely increase with energy beyond the mobility edge, avalanche becomes a possibility. This argument neglects the possibility of the existence of localized phonons, which could have energies significantly greater than the maximum optical-phonon energy of the corresponding crystal, and thus could take away much of the excess energy of the hot electrons, eventually thermalizing the distribution. No experimental evidence is available which bears on this possibility.

Zener breakdown can, in all likelihood, be eliminated, either on the basis that fields of 10^5 V/cm cannot yield a tunneling probability through a 0.5 eV barrier sufficiently large to overcome the recombination rate, or because of the fact that the large negative-resistance regions observed are incompatible with the necessity of maintaining the tunneling rate.

Ferroelectric-type phase transitions have been proposed by Drake et al.,[907-909] but these are not as likely in the primarily covalent chalcogenide glasses as in the ionic transition-metal-oxide glasses. The strength and directionality of the covalent bonds together with the small ionic charge in the chalcogenides[539] make it improbable that a phase transition can be induced at fields of only 10^5 V/cm. However, there is no real quantitative test of this model at present.

Hed and Freud[910] suggested that switching could be induced by a decoupling of small polarons by field-generated carriers. This model does not appear to be applicable to the chalcogenide glasses, in which conduction is believed to be primarily band-like, rather than by hopping. A similar objection can be raised against the model of Bagley,[911] who proposed that switching results from a sharp increase in mobility with field, analogous to that obtained in small-polaron theory. Mattis[912] suggested a model that is a version of a nonequilibrium Anderson transition. In this approach, excess carriers generated by Zener tunneling produce an effective reduction of the mobility gap, the decrease arising from the delocalization of states near the equilibrium mobility gap due to screening by the excess carriers. This model exhibits a sharp conductivity discontinuity,[904] at which point the mobility gap collapses and the material becomes essentially metallic. The model has difficulty in explaining the independence of the holding voltage on thickness, since the consequent lack of field in the bulk should then bring about a return to equilibrium.

The most popular model for switching in chalcogenide glasses is that of thermal runaway.[856,857,871,913-925] Electrothermal effects in solids have had a long and glorious history[926-934] dating back to the eighteenth century, and all modern calculations risk the discovery that their results were well known during the Franco-Prussian War. There is, in fact, a great deal of appeal to the thermal-runaway model it is a quantitative model, which, in principle, can account for the negative resistance, the preswitching delay time, and the filament formation.

The fact that a thermal model for switching can be made quantitative suggests the hope that results can be obtained and tested experimentally. In fact, this is a rather naive view of the scope of the problem, as we shall now see. A proper treatment begins with the linear flow equation,

$$\underset{\sim}{J} = \sigma \left(\underset{\sim}{F} - \frac{\nabla\mu}{e}\right) + S\sigma\ \nabla T$$

$$\underset{\sim}{J}_Q = ST\underset{\sim}{J} - \kappa\nabla T \quad ,$$

where **F** is the applied electric field, μ is the electrochemical potential, $\mathbf{J}_Q$ is the heat current density, **J** is the electrical current density, S is the thermoelectric power, and κ is the thermal conductivity. In addition, we must impose conservation of charge,

$$\nabla\cdot\underset{\sim}{J} = -\frac{d\rho}{dt} \quad ,$$

where ρ is the charge density, conservation of energy,

$$\nabla \cdot \underset{\sim}{J}_Q = \underset{\sim}{J} \cdot \left(\underset{\sim}{F} - \frac{\nabla\mu}{e}\right) - T\frac{dS}{dT}\underset{\sim}{J} \cdot \nabla T = c_v \frac{dT}{dt} \quad ,$$

and the constitutive relations, S(T), κ(T), and σ(T) $= \sigma_o \exp(-E_A/kT)$. Furthermore, we must specify the geometry, boundary conditions, and initial conditions. Since the problem is inherently inhomogeneous, all parameters have a spatial dependence.

It is quite evident that the problem cannot be

solved without several approximations. It is always assumed that the terms involving $\nabla\mu$ and S can be neglected. These assumptions can be tested after the equations are solved. The time-dependent solution is much more difficult to find than the steady-state result, and only the latter is investigated. In a more sophisticated approach, the steady-state solution can be used as the infinite-time boundary condition to obtain the time dependence, but this is a formidable problem. Typically, a cylindrical geometry is used, to reduce the dimensionality somewhat, and the isotropy of the amorphous semiconductor parameters provides a further simplification. The final set of equations is then

$$\kappa\nabla^2 T(\underset{\sim}{r}) + \frac{J^2(\underset{\sim}{r},T)}{\sigma_o} e^{E_A/kT} = 0 \tag{43}$$

$$\nabla \cdot \underset{\sim}{J}(\underset{\sim}{r},T) = 0 \quad . \tag{44}$$

Another important assumption almost always made is that the electrodes act as perfect heat sinks, thus keeping the electrode-semiconductor interfaces at ambient temperature. This is probably a reasonable assumption.[935] Since the thermal conductivity of a typical chalcogenide glass is 0.002 watt/cmK[633] as compared with 2 W/cmK for Mo electrodes, the electrodes conduct heat away 1000 times faster than the glass. Thus, an increase in electrode temperature of 1K is equivalent to an increase in temperature of the amorphous material of 1000K, if both materials are kept equally far from a heat sink. Recent thermocouple measurements[936] have indicated only a negligible heating of the electrodes unless the device is operating far above the holding current.

The importance of the assumption that the electrodes are infinite heat sinks has been recently emphasized by Keyes,[937] who showed in a very simple manner that in one dimension, this restriction prevents any negative-resistance region from occurring in a pure thermal model. Since the heat-balance equation in one dimension is

$$-\kappa \frac{d^2T}{dx^2} = \frac{J^2}{\sigma} \quad ,$$

and the current density is related to the electrostatic potential by

$$J = \sigma \frac{dV}{dx} \quad ,$$

then

$$\kappa \frac{d^2T}{dx^2} = -J\frac{dV}{dx} \quad . \tag{45}$$

But since J and κ are independent of x, Equation 45 can be integrated to give

$$\kappa \frac{dT}{dx} = -JV$$
$$= -\frac{1}{2}\sigma\frac{dV^2}{dx} \quad . \tag{46}$$

Thus:

$$V^2 = -2\int_{T_m}^{T_e} dT \frac{\kappa}{\sigma} \tag{47}$$

where T_m is the maximum temperature at the center, T_e is the temperature at the electrode, and V is half the voltage applied across the device. Differentiating Equation 47, we find

$$\frac{dV^2}{dI} = 2\kappa\left[\frac{1}{\sigma(T_m)}\frac{dT_m}{dI} - \frac{1}{\sigma(T_e)}\frac{dT_e}{dI}\right] \quad . \tag{48}$$

If the electrodes are infinite heat sinks, dT_e/dI = O, and dV^2/dI must always be positive. Thus, no negative-resistance region can exist.

Most explicit solutions of the thermal problem make one or more of the following assumptions in addition to those discussed previously:

a. The temperature distribution throughout the material is uniform.[915,921-924]

b. The electric field is constant throughout the material.[917-924,938]

c. The heat equation is solved in one dimension, with heat flow only perpendicular to the current flow.[917-920,942]

d. The heat equation is solved in one dimension, with heat flow only parallel to the current flow.[915,916,921-924,938-941]

e. The functional dependence of the conductivity on temperature is of the form $\sigma = \sigma_{oo}$ exp $[\beta$ (T-T)] or $\sigma = AT^n$, rather than the experimentally observed form.[923,938-941,943]

It turns out that any of these assumptions is sufficient to invalidate even the qualitative features of the results of the calculation.

Kaplan and Adler[944] numerically solved the steady-state problem, Equations 43 and 44 without any assumptions (a) through (e), and found no negative-resistance region occurs and no high-current filaments form, provided no electronic effects are included. These results are shown for typical values of the parameters in the curve labeled STM in Figure 46. Thus, even in three dimensions, no pure thermal switching can take place when the electrodes are infinite heat sinks. On the other hand, when a virtual-electrode model, in which some electronic space-charge effects are explicitly introduced, is employed, both a CCNR region and a sharp high-current filament are found. The I-V characteristic is the curve labeled VEM in Figure 46. The virtual-electrode model assumes that the equipotential plane near each electrode is displaced a small distance inside the active material. Such a model can be achieved physically if either ordinary space-charge injection exists or sufficiently narrow Schottky barriers form.

Since the complete results differ from those obtained using assumptions (a) through (e) it is clear that these assumptions are inadequate. Both assumptions (a) and (c) lead to a constant temperature parallel to the direction of current flow, thus allowing the semiconductor-electrode interfaces to heat. It is this heating that leads to the appearance of a negative-resistance region. Assumption (b) is simply not self-consistent, since the spatial variation of the temperature across the device forces a sharp increase in field near the electrodes. Assumptions (d) and (e) do not lead to spurious negative resistance, but do affect the nature of the solutions. Assumption (d) results in much too large values of the maximum temperature at the center of the device and assumption (e) leads to a spurious rapid increase of current with voltage at a critical field, resulting in a characteristic similar to homogeneous Zener breakdown. The physical reason that no negative resistance is found in the pure thermal model with infinite heat sinking is that the region near the electrodes act as current-density-limiting resistances. The virtual-electrode model allows CCNR by providing a mechanism by which these resistances are effectively short-circuited. Explicitly introducing a field dependence into the conductivity, $\sigma = \sigma_o \exp(-E_A/kT) \exp(F/F_o)$, as has been observed,[860] also accomplishes a similar short circuiting, and produces negative resistance with or without virtual electrodes.[944]

If such an electro-thermal model as just described can produce a CCNR, we must still compare its predictions with the experimental results before deciding on its applicability. Electronic processes require a threshold field essentially independent of the thickness of the film. On the other hand, thermal or electro-thermal CCNR generally leads to a gradual decrease of threshold field with increasing film thickness. A revealing

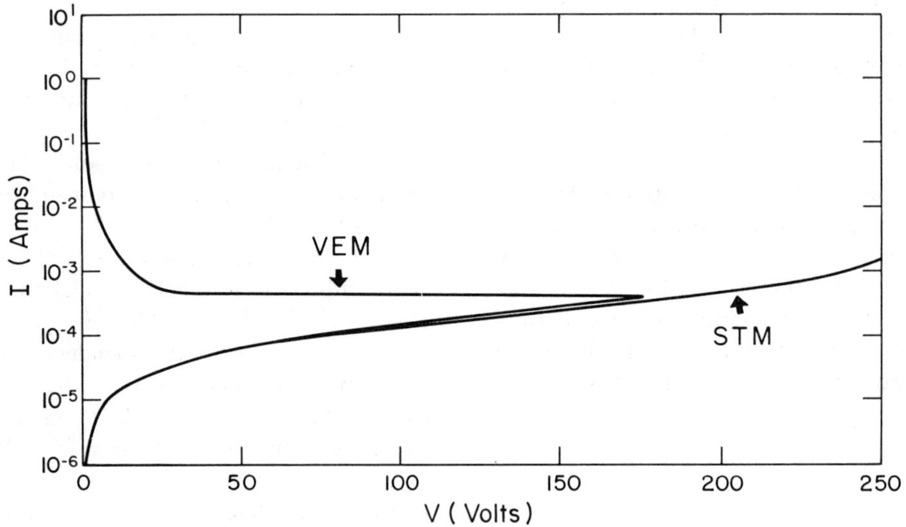

FIGURE 46. I-V characteristics for standard thermal model (STM) and virtual-electrode (VEM) for E_A = 0.5 eV, κ = 0.002 watts 1 m-k, T_o = 300K, σ_o - 10^3 Ω^{-1} cm^{-1} for a cylindrical sample of 100 μ radius and 10 μ thickness (Kaplan and Adler[944]).

experiment has been performed by Kolomiets et al.,[945] who measured the threshold field of a Te-As-Ge-Si glass as a function of thickness from 0.3 μ to 1000 μ. The results are shown in Figure 47. For films thinner than 5μ, the threshold field was found to be independent of thickness and only slightly dependent on temperature. On the other hand, films thicker than 50μ exhibited a threshold field that varied strongly with temperature and decreased with increasing thickness. Kolomiets et al. concluded that switching was thermally initiated in the thicker films, but was of electronic nature in the thinner films.

Experimental results, such as the thickness dependence of the threshold field, vary considerably with composition. Stocker et al.[938] found a thickness-independent region in $Te_{50}As_{30}Ge_{20}$ only up to about 0.5 μ; the threshold field decreased with increasing thickness for thicker films. There are also several observations of hot filaments in the conducting state of particular compositions. Pearson and Miller[856] filmed evidence for a liquid phase in a 10 μ film of Te_2As_2Se. Weirauch[946] studied a 0.2 cm bulk sample of amorphous $Se_{50}Ge_{30}As_{20}$ with an infrared viewer and observed a bright filament whose emitted radiation was equivalent to a black body at 900K when the sample was switched at a 550K ambient. Gunthersdorfer[914] observed a molten area in a 1 μ film of a Te-As-Ge-Si glass by means of a scanning electron microscope and calculated that the temperature had risen about 500K. Armitage et al.[865] also investigated a 1 μ sample of $Te_{48}As_{30}Ge_{10}Si_{12}$ with a scanning electron microscope, and observed a melting of the gold electrodes. However, none of these experiments indicate the mechanism of the actual switching, and the variation in heating effects arising from variables such as the self-capacitance of the device, the load resistance, and the switching repetition rate make it extremely difficult to generalize this type of observation.

There are several experimental results that appear to be in conflict with a thermal model for switching in the thinner chalcogenide-glass films. (1) Calculations[944] indicate that the threshold fields in a 1 μ film is of the order of 10^6 V/cm, much larger than the experimentally observed values. (2) Pulse measurements suggest that the effects of Joule heating during the delay time are insignificant,[834] even though strong non-ohmic conduction is present. This indicates the apparent

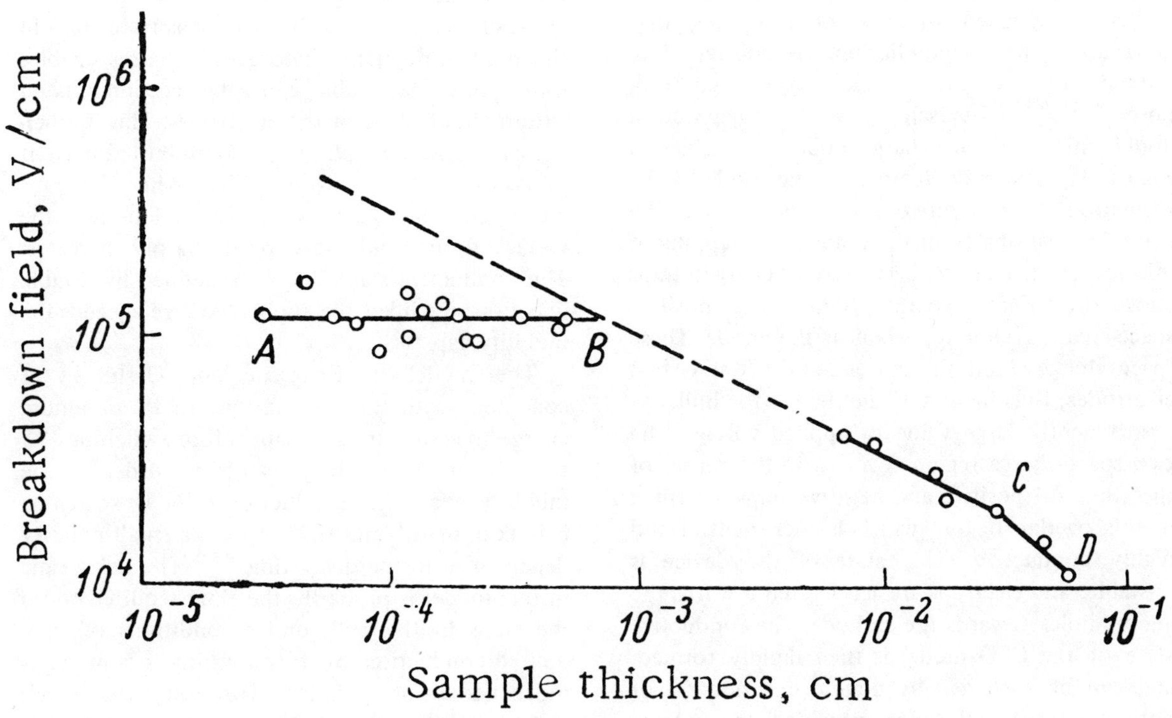

FIGURE 47. Dependence of threshold field on sample thickness for a Te-As-Ge-Si glass (Kolomiets et al.[945])

field dependence of the conductivity is an electronic effect, probably due to Poole-Frenkel emission. (3) The switching time, of the order of 10^{-10} sec, is too fast to explain the temperature rise necessary to produce the ON-state conductivity,[834] although the possibility of a conductivity anomaly can modify this objection.[947] (4) It is difficult, given the I-V characteristics resulting from thermal switching models, to account for the observation that the holding voltage is independent of the load resistance. (5) Thermal theories suggest that the delay time reflects the time necessary to heat the material to the runaway point, but experiments indicate that switching occurs after a fixed charge is injected into the material,[948] and not a fixed power. However, there is some evidence that this is in accord with electrothermal predictions.[949] (6) Henisch et al.[950-952] have observed a small, but definite, polarity dependence to both the delay time and the threshold voltage, a result that is difficult to explain on the basis of a thermal model. Although *ad hoc* extensions of thermal models appear to be able to bring the predictions into agreement with the experimental results, in the absence of explicit quantitative calculations it is impossible to make any definite statements as to whether switching is primarily thermally or electronically initiated.

Several detailed models for switching that depend on the nonequilibrium conducting state implicit in the CFO model have been proposed.[953-957] Henisch et al.[954] suggested a double-injection space-charge model, as outlined in Figure 48. The main feature of the model is the formation of a negative space charge near the cathode, resulting from the immediate trapping of injected electrodes by the positively charged traps above the Fermi energy. Similarly a positive space-charge region appears near the anode. These space charges limit the current and act as virtual electrodes, thus increasing the field in the bulk. At a sufficiently large value of applied voltage, the two space-charge regions overlap in the center of the film. All positive and negative traps are filled in this overlap region, which is then neutral and highly conductive. This state of the device is unstable, and electrons are accelerated towards the anode, holes towards the cathode. The conductive state of the CFO model is then rapidly formed, and can be sustained by double injection[958] of both electrons and holes, provided the voltage across the device is larger than the mobility gap. Since the multicomponent chalcogenide glasses form ohmic contacts, double injection is possible without a distinct dependence on the electrodes.

In the model of Henisch et al.,[954] the delay time is the time required for the two space-charge regions to overlap. It might then be expected that the delay time should increase strikingly if the polarity of the voltage pulse is reversed prior to switching. Such is not the case at room temperature,[913,959] although there is a slight increase of the delay time after the reversal.[950] This increase in delay time becomes substantial at low temperatures,[952] indicating that space-charge effects become important in that temperature range. At room temperature, there are sufficient phonons present for the trapped carriers to equilibrate too rapidly for large space-charge regions to occur. However, the model of Henisch et al.[954] may well be applicable to switching at 77K.

Fritzsche and Ovshinsky[953] proposed a somewhat different model, which is sketched in Figure 49. They assume that the non-equilibrium carriers generated by the field cause two depletion regions to form, one for electrons near the cathode and one for holes near the anode. These depletion regions arise because the electrodes cannot supply a sufficient carrier flux to maintain the excess current. The resulting Schottky barriers grow with increasing voltage, but the large density of traps in the multicomponent chalcogenide glasses enables the entire space charge to be accommodated within about 30 Å of the electrodes. Thus tunneling can easily take place, and trap-limited current proceeds until all traps are filled. When the traps are filled, the state is highly conductive, the voltage drops, and the current sharply increases. The conductive state is again sustained by double injection, provided the applied voltage exceeds the mobility gap.

The model of Fritzsche and Ovshinsky is consistent with the observations that a minimum charge injection must occur before switching,[948] that the holding voltage is of the order of the mobility edge, that conduction prior to switching is bulk-limited,[1] and that there is a small polarity dependence to the delay time.[950] The delay time in this model is primarily the time required to fill the traps in the bulk under conditions of trap-limited conduction. Such trap-filling is completely independent of polarity. However, the much shorter time required to set up the Schottky barriers is a strong function of the sign of the

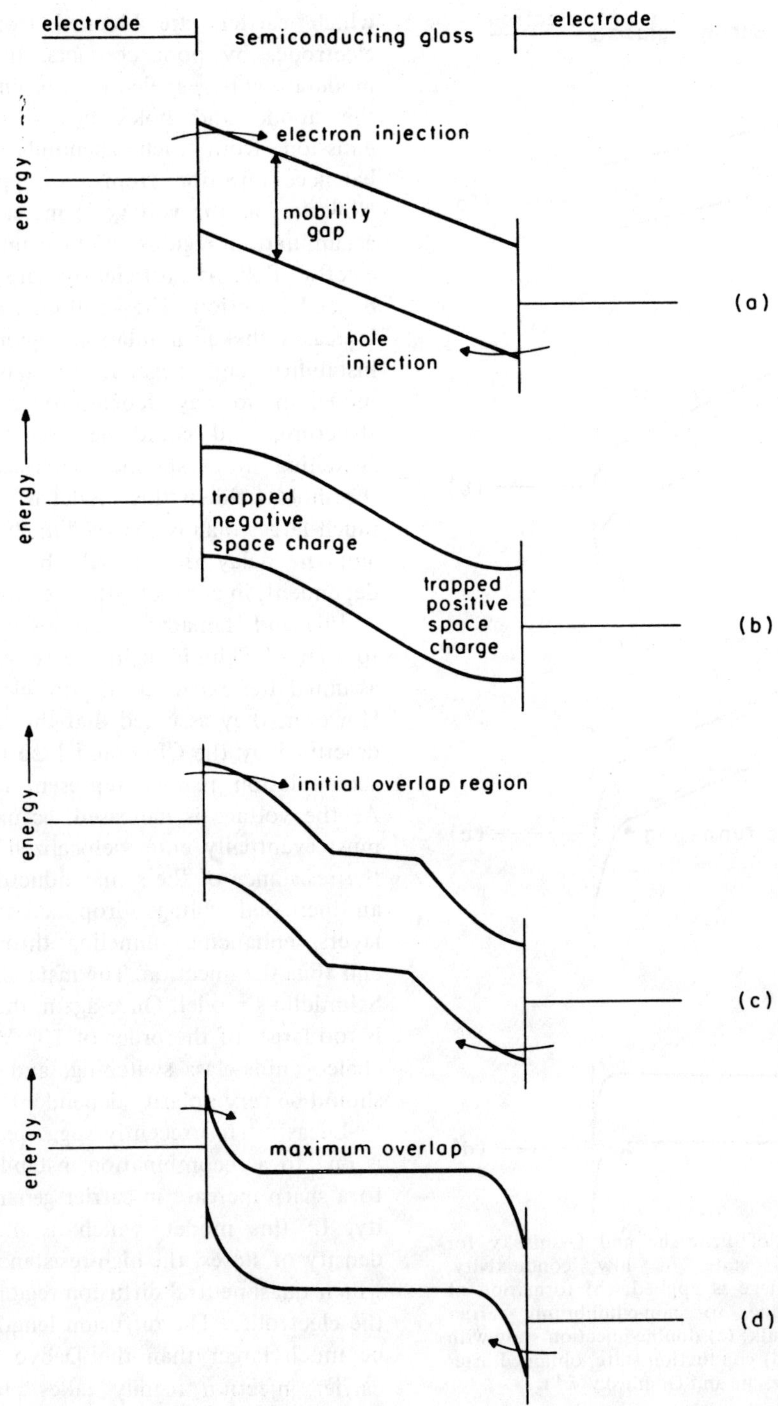

FIGURE 48. Space-charge double injection model for switching: (a) initial state of low conductivity, immediately after voltage is applied; (b) formation of space charges and space-charge-limited current; (c) unstable situation resulting from overlap of the positive and negative space-charge regions; (d) conducting state, maintained by double injection of electrons and holes (Henisch et al.[954]).

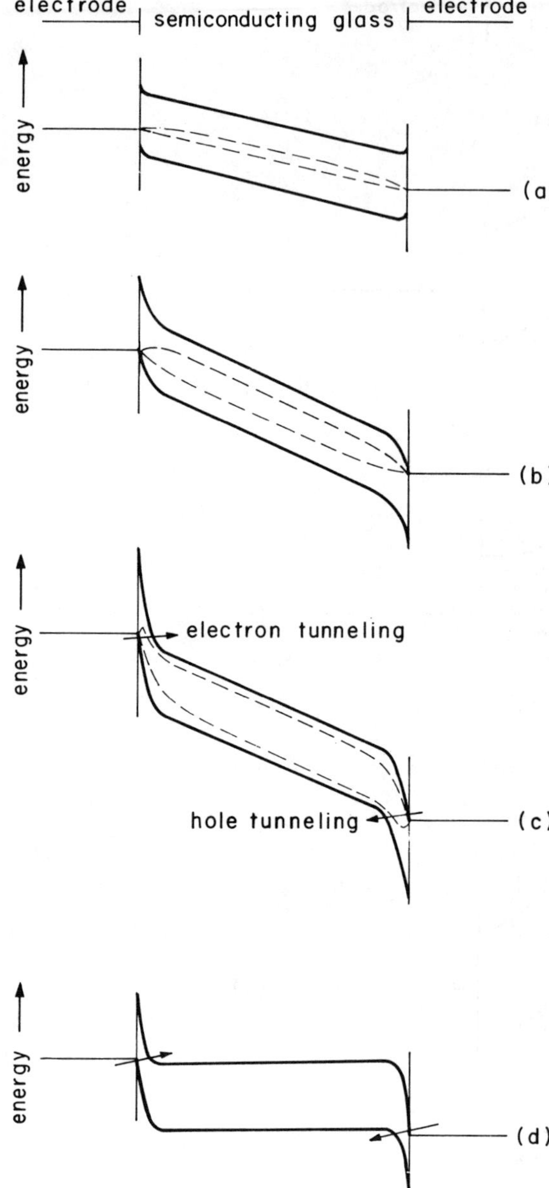

FIGURE 49. Model of Fritzsche and Ovshinsky for switching: (a) initial state of low conductivity, immediately after voltage is applied; (b) formation of Schottky barriers, due to non-equilibrium carrier concentration in the bulk; (c) double-injection state with trap-limited current; (d) conduction state, obtained after all traps are filled (Fritzsche and Ovshinsky[953]).

voltage, thus giving a small polarity dependence to the total delay time.

Schmidlin[955] noted that I-V characteristics similar to those observed in switching can be obtained with a model of field-induced double injection, sustained by accumulation regions in which carriers are blocked from entering the electrodes by poor contacts. In this model, at moderate voltages, electrons begin to pile up near the anode and holes near the cathode until emission from each accumulation region just balances injection from the opposite electrode. However, as the voltage is increased further, the accumulation regions increase until the resultant electric field is sufficiently large to induce enhanced injection. The additional injection further increases the accumulation region, leading to an instability and negative resistance. Schmidlin's model in no way depends on a CFO-type band structure, and could be used to account for switching in crystalline materials. However, the threshold field in this model is about 10^6 V/cm, much larger than is observed in chalcogenide films, and the delay time should be strongly polarity dependent, in conflict with the observations.

Iida and Hamada[956] proposed a model similar to that of Schmidlin in the sense that they also assumed the existence of poor electrical contacts. However, they assumed that the semiconductor is described by the CFO model, so that initially the accumulation regions represent trapped carriers. As the voltage is increased, accumulated carriers must eventually enter delocalized states, lowering the resistance of the semiconductor. This leads to an increased voltage drop across the insulating layers, enhancing tunneling through the barrier and thus the injection. The instability follows as in Schmidlin's model. Once again, the threshold field is too large, of the order of 10^6 V/cm, to explain chalcogenide-glass switching, and the delay time should be very polarity-dependent.

Lucas[957] has recently suggested that switching is due to a recombination instability, rather than to a sharp increase in carrier generation or mobility. In this model, which assumes a CFO-type density of states, the high-resistance state is one in which quasi-neutral diffusion regions build up near the electrodes. The diffusion length is assumed to be much larger than the Debye length, so that carrier injection readily takes place and nearly equal densities of electrons and holes are trapped in the bulk, primarily in the charged traps near the Fermi energy. As the charged traps fill up, they become neutral, sharply decreasing their cross section, and thus increasing the mean diffusion length. Lucas calculated from the $\mu\tau$ product observed in photoconductivity experiments[672] that the diffusion length is initially of the order of

0.1 μ. Since neutral traps are only about 1% as efficient as charged traps,[960] the diffusion length increases to approximately 10 μ after the traps are filled. Since this is larger than the thickness for thinner films, the conductivity sharply increases and the voltage drops to zero across the bulk. This model is very similar to the one of Fritzsche and Ovshinsky,[953] and has the same advantages insofar as agreement with experiment is concerned. The main difference between the two models is the assumption of a quasi-neutral diffusion region by Lucas. Because of this assumption, the trap filling need only take place over a single diffusion length to induce switching.

It should be noted that CCNR triggered by double injection is commonly observed in crystalline material, such as doped Ge,[961,962] and that high-current filaments have been studied in detail.[963,964] These current filaments have been shown to have electronic rather than thermal origin.[965]

A few attempts to characterize the nucleation and growth of the high-current filament have been made.[966,967] Chakraverty[966] considered the nucleation kinetics of a filament in which the low resistance is due to a field-induced collapse of the mobility gap. In this model, the delay time is proportional to exp $[K(V-V_o)^{-1}]$, where V_o is the voltage necessary to reduce the mobility gap to zero, and K is a constant independent of field but inversely proportional to temperature. This functional dependence of delay time on applied voltage is not found in chalcogenide glasses,[834] but has been observed in crystalline Fe_3O_4.[884] Homma[967] has suggested that a high-conductivity plasma in which all traps have been filled by injected carriers, forms a filament near one electrode. This low resistance filament then propagates to the other electrode, thus inducing switching. The delay time in this model is the time necessary for the plasma to complete a path between the electrodes. An objection to this model is that the resistance of the device should decrease continuously as the plasma propagates, due to the shorter path between the front of the plasma and the second electrode. No such resistance drop is observed experimentally in pulse experiments.

To summarize, it appears that switching in thicker films is predominantly thermal in nature, although some electronic effects are necessary. In thinner films, of the order of 1 μ, for at least a large class of compositions and geometries, switching is most likely electronic in origin, although the details of the mechanism are far from clear at this time.

B. Voltage-Controlled Negative Resistance

Voltage-controlled negative resistance (VCNR) has been observed often, primarily in amorphous oxides. Kreynina et al.[968-970] were apparently the first to observe VCNR in amorphous materials, in 600 Å - 1.5μ anodic films of Al_2O_3. Hickmott[971-973] reported irreversible VCNR in 150-1000 Å anodic films of Al_2O_3, Ta_2O_5, ZrO_2, TiO_2, and SiO. The breakdown field was of the order of 10^6 V/cm, and the switching time was less than 0.5 sec. Hickmott proposed that electrons were tunneling in non-equilibrium trapping levels above the Fermi energy. The negative resistance then could represent field-induced equilibration of the trapped electrons.

A systematic investigation of VCNR in amorphous SiO was carried out by Simmons and Verderber.[974-976] They suggested that the Fermi energy of the film is pinned in a narrow impurity band. In low fields, conduction is primarily by hopping in the impurity band, but at high voltages space charge can fill the impurity band, decreasing the current.

Other models for VCNR have been proposed. Barriac et al.[977-979] suggested that the current is initially due to space-charge-limited ionic conduction, but as the field increases, Schottky emission leads to a trapping of the electronic carriers by the ions, thus reducing the current. Dearnaley[980,981] proposed that high-conductivity filaments are formed, but these filaments are thermally fractured at high fields, leading to a drop in the current.

VCNR occurs only after a forming process,[974] which can be accomplished by cycling the sample in a field of a given polarity. The resistance of the film decreases by a factor of about 10^8 during formation. The forming field is of the order of 10^6 V/cm.[976] The models described previously to account for VCNR all suggest different explanations for the forming process. Hickmott[973] proposed that the high field ionizes impurities by a Poole-Frenkel emission, thus setting up an impurity band. Simmons and Verderber[974] interpreted forming as the injection of positive ions from the anode into the oxide, producing impurity levels near the Fermi energy. Barriac et al.[979] also suggested the injection of positive ions into the oxide, due to local melting near the metaloxide interfaces. Dearnaley et al.[982] proposed that formation was a structural change in the oxide, resulting in a conducting filament. It is difficult to differentiate between these models on the basis of the present experimental data.

C. Memory Phenomena

Memory switching has been observed in an extremely large class of materials, including chalcogenide glasses[1,851,853,983,984] and oxide glasses.[985-989] It is important to differentiate an intrinsic switching process, due to changes within the semiconductor itself, from artifact switching, due to the formation of a metallic bridge between the electrodes and the subsequent thermal rupture of such bridges. Artifact switching has been investigated in such materials as mica, Saran Wrap,[990] and Mylar,[991] and doubtless could be briefly observed in a sheet of the New York Times. The most important difference between the threshold switching described in subsection A and memory switching is that films which exhibit memory must have two distinct *equilibrium* states. Because the conducting state can be studied and characterized at equilibrium, memory switching is at present better understood than threshold switching.

A typical memory switch behaves initially just like a threshold switch. Conductivity is very low until a given voltage is exceeded. After a delay time, there is a very rapid switching to a highly conductive state. If the current is quickly reduced below a critical value, the device will switch back to the resistive state, exactly like a threshold switch. However, if the device is held in the conductive state for the order of 10^{-3} sec, it will remain highly conducting, even after the external field is removed. Once locked in the ON state, the material can be returned to the original OFF state by applying a short, intense current pulse of either polarity.

In the artifact memory discussed previously, the ON state is just electrode material sputtered through the film or occasionally a filament of reduced material, e.g., a Cu filament embedded in CuO.[990] The return to the OFF state is just a fuse effect – the intense current pulse burns out the filament. In true memory devices, the ON state has been shown to be due to a phase separation and crystallization of the amorphous material.[992-995] In fact, the difference between switching-type and memory-type glasses is that crystallization occurs much more readily in the latter than in the former.[996] Switching material may also phase separate, but both phases generally remain amorphous.[997] A typical example of memory-type material is amorphous Te_{50} As_{40} Ge_{10}, which phase separates and crystallizes near 550K.[998] The phase responsible for the high conductivity in this case has been shown to be degenerate crystalline As_2Te_3. A similar material, amorphous As_{55} Te_{35} Ge_{10}, has been studied by Uttecht et al.,[999] who observed first a threshold switching and then a memory effect on the surface of a bulk sample, with the electrodes about 600 μ apart. After switching, a filament began growing from the positive electrode. If the applied field was removed before the filament reached the negative electrode, the material switched off. However, once the filament completed a path between the electrodes, the memory state was achieved. If the polarity of the applied field was reversed before the filament was complete, the existing filament rapidly disappeared, and a new one began growing out the opposite (now positive) electrode. Microprobe measurements showed that inside the filament, the Te concentration increased 38%, while the As concentration decreased 41% and the Ge concentration decreased 47%. The composition inside the filament was very close to that of heavily Ge-doped As_2Te_3. The filament growth in this case is very suggestive of electromigration, particularly because of the polarity effects.

Not all memory devices exhibit such polarity dependence. In particular, the amorphous alloy Te_{81} Ge_{15} As_4 and similar compounds in which the As is replaced by other elements have been extensively studied. The key to memory behavior in this material is the calorimetry experiments of Fritzsche and Ovshinsky.[992] Figure 50 shows results of a different thermal analysis of bulk amorphous Te_{82} Ge_{16} Sb_2. In such an experiment, the material is heated at a constant rate, and the signal obtained measures any change in the rate of heat absorbed or given off when compared to a standard of constant heat capacity. Thus, phase transitions in the material being investigated can be identified. Figure 50 (a) shows the result when the glass is heated from room temperature to 475°C. At about 125°C, a drop in the signal indicates the glass transition temperature, at which point the viscosity of the material decreases sharply. With continued heating to approximately

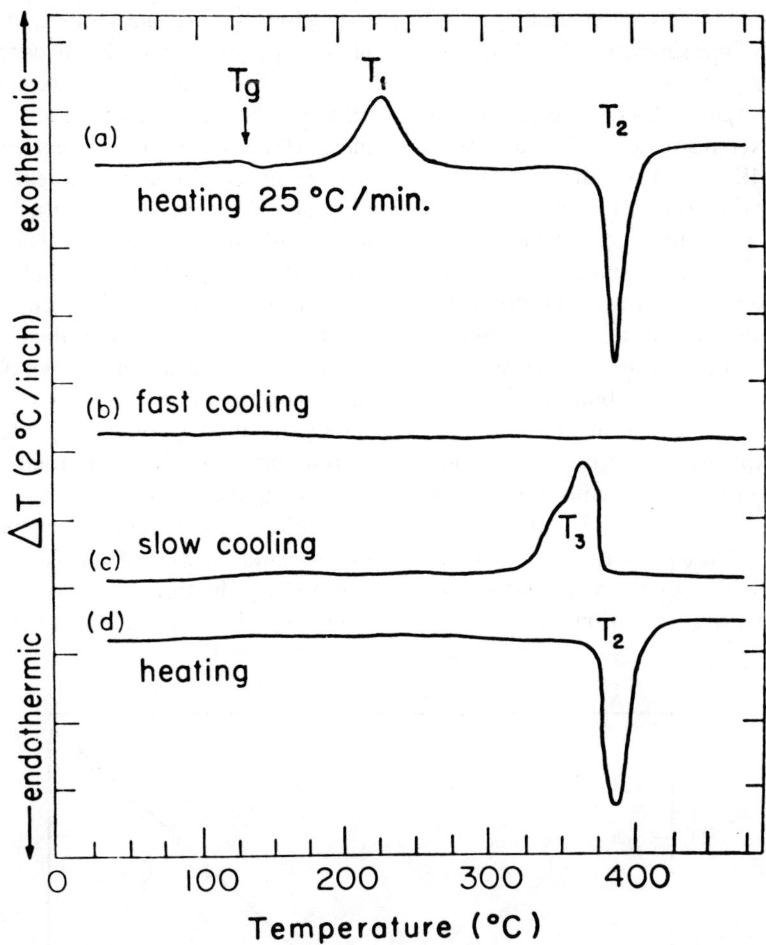

FIGURE 50. Differential thermal analysis of a memory-type chalcogenide alloy, Ge_{16} Te_{82} Sb_2: (a) heating from the amorphous phase; (b) rapid cooling from the liquid phase; (c) slow cooling from the liquid phase: (d) heating from the crystallized phase. (Fritzsche and Ovshinsky[992])

225°C, a large increase in the signal occurs, indicating the occurrence of an exothermic transition. This transition represents a phase separation into Te-rich and GeTe-rich regions. Since amorphous Te crystallizes above 10°C,[180] the Te is crystalline. Indeed, x-ray analysis has shown the existence of polycrystalline Te above the transition.[547]

When the material is cooled either slowly or rapidly from the region between the exothermic peak and the large endothermic peak near 375°C, the conducting state of the material is obtained. The endothermic peak represents melting of the crystallized material. If the material is cooled rapidly from the liquid phase, Figure 50 (b) is obtained. No transitions at all are found, indicating a return to the glassy state by freezing in the disorder of the liquid. The resulting material has high resistance.[539] On the other hand, Figure 50 (c) shows that a slow cooling from the liquid phase produces an exothermic transition, corresponding to a solidification to the highly conducting crystallized state. A reheating of this material, as is shown in Figure 50 (d), shows only a melting signal.

Nuclear magnetic resonance studies[616] indicate that the crystallized material primarily consists of highly doped Te. The presence of crystalline Te has been observed by electron diffraction and scanning electron microscopy in phase-separated chalcogenide glasses.[1000] In addition, electrical transport studies of Te_{81} Ge_{15} As_4 in the conducting state[539] have shown properties very similar to those of degenerate, crystalline As-doped Te. Figure 51 shows the result of a four-probe resistivity measurement from 4K to 300K. Above 20K, resistivity increases essentially linearly with temperature, although a slight break occurs at 180K. No resistance minimum was observed down to 4K, indicating the material is degenerate rather than extrinsic. Hall effect measurements indicate a carrier concentration of approximately 10^{20} cm^{-3}, independent of temperature. The room-temperature mobility is 85 cm^2/V-sec. The degenerate nature of the material is rather strikingly brought home by the appearance of superconductivity below about 0.8 K.[1001]

Electrical memory switching in this class of materials can be explained in terms of the observed phase transitions.[539] The amorphous film is subjected to a high field which induces a non-equilibrium conducting state. The concomitant Joule heating raises the temperature of the material above the glass transition temperature but below the melting temperature. In this region, phase separation and crystallization of Te can take place. The crystallization takes place in about a millisecond, much more rapidly than in the bulk material, even considering that only the volume of the small conducting filament need be transformed. The reason for this is that, in the non-equilibrium conducting state, many of the covalent bonds are broken and the phase separation is greatly accelerated. Electric-field induced crystallization has been studied in anodic Ta_2O_5 films,[1002] and is a well known phenomenon. The sharp current pulse which switches the device off simply melts the conducting filament. Since the surrounding material is essentially at room temperature, a rapid cooling assures glass formation.

Feinleib and Ovshinsky[1003] performed optical experiments on bulk $Te_{81}Ge_{15}As_4$, and found that the conducting state exhibited a reflectivity essentially the same as that of polycrystalline Te.

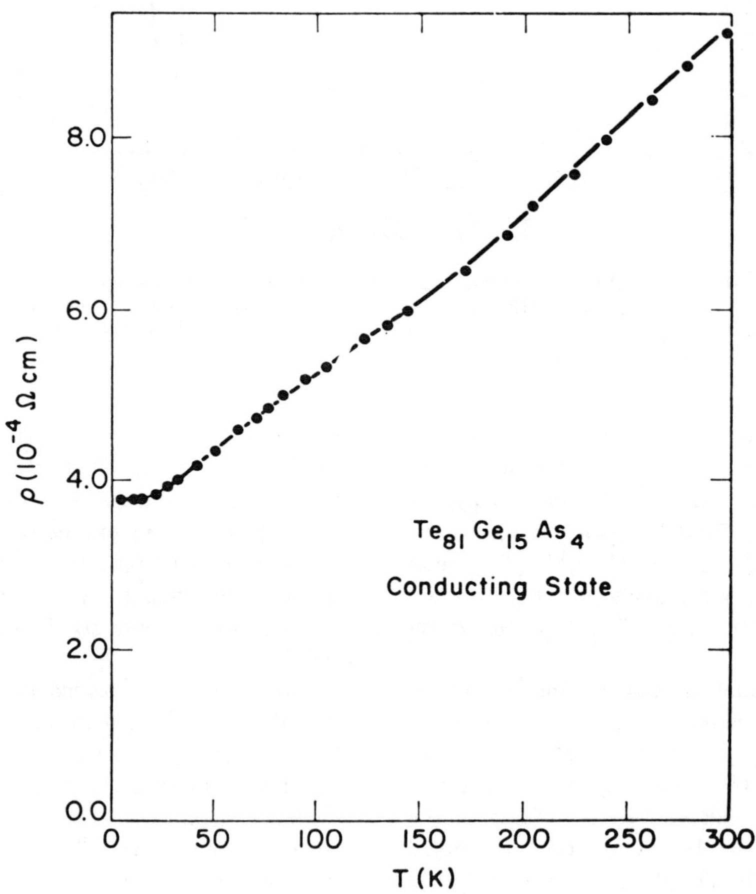

FIGURE 51. Resistivity of the conducting state of $Te_{81}Ge_{15}As_4$ as a function of temperature (Adler et al.[539]).

Comparative reflectance studies on thermally cycled material are shown in Figure 52. The reflectivity of the conducting state below 4 eV is approximately 50% larger than that in the nonconducting state, independent of photon energy and previous thermal history.

Evans et al.[1004] found that a flash zenon lamp could be used to switch amorphous Te_{85} Ge_{15} to the conducting state provided the flash time exceeded 3 msec. Feinleib et al.[1005-1007] focused argon-ion laser radiation of 5145 Å wavelength on a 5μ spot of amorphous Te_{81} Ge_{15} Sb_2 S_2, and found that a 1 μ sec pulse of 100 mW induced crystallization. A pulse of approximately the same power and duration then revitrified the material. The extremely fast speed used to crystallize the spot implies that a photocrystallization effect was operative. Since the photon absorption necessarily creates broken bonds, a greatly enhanced crystallization growth rate can be expected. However, the fact that the same energy pulse can crystallize and revitrify the material is somewhat surprising. A model for this effect is shown in Figure 53. The optimum condition for crystallizing the amorphous material at ambient temperature T_A occurs when the pulse intensity and duration are such as to bring the material to the temperature at which the crystallization rate is a maximum, without exceeding the melting temperature T_M. This is shown in curve A. The material is then held at temperatures between the glass transition temperature T_G and the melting temperature for a time t_1. During this time, thermal growth of crystallites can occur. However, during a fraction t_2 of this time, the light is on, and photocrystallization greatly enhances the growth rate. Since the nonequilibrium state is preserved for a time after the light is turned off,[666] the enhancement of crystallization can continue even in the remaining part of t_1. Providing T_M is not reached, the material will remain crystallized after cooling to T_A.

In the reverse process, curve C, the crystallized material is also heated by the absorption of light energy. But since the absorption constant of the crystallized material is larger than that of the amorphous material at 5145 Å,[181] and since, in addition, the crystallized material has a lower heat capacity than the amorphous material between T_G and T_M,[633] the crystallized material will reach a higher temperature with the same or even less laser energy than will the same volume of the glass. If

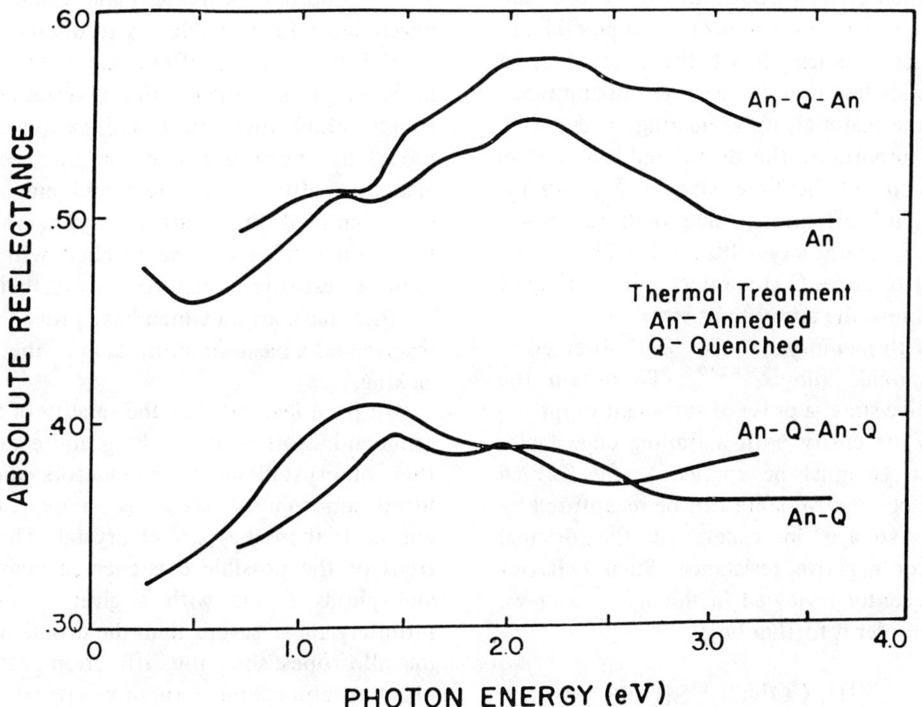

FIGURE 52. Reflectance of bulk samples of Ge_{15} Te_{81} As_4. Quenched material is in the amorphous state, annealed material is in the crystalline state (Feinleib and Ovshinsky[1003]).

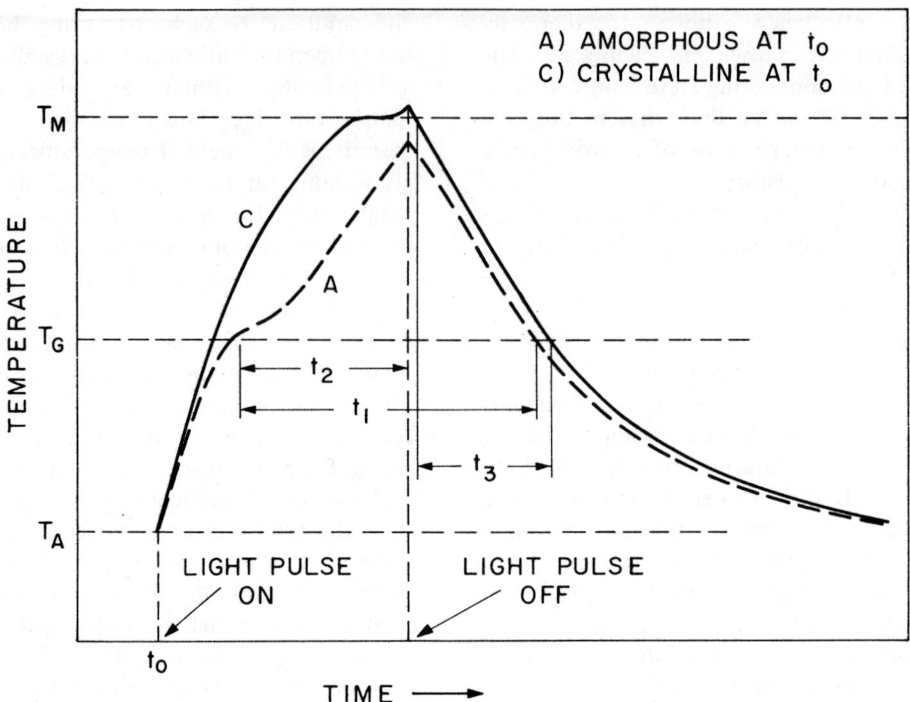

FIGURE 53. Temperature-time profile of the active material illuminated by a square light pulse energy above both amorphous and crystalline band gaps (Feinleib[1005]).

the maximum temperature reached exceeds T_M, the material melts. After the light is turned off, the liquid then cools through the range between T_M and T_G, in which crystallization is possible, in time t_3. Since t_3 is less than t, the material, upon cooling, spends less time in the crystallizing range than does the material, upon heating, in curve A. But, more important, the disordered material in curve C spends all the time between T_M and T_G while the light is off, so that the growth rate is not enhanced by photocrystallization. Thus, the cooling part of curve C guarantees a quenching of the material into the amorphous state.

VCNR with memory has also been observed in amorphous oxide films.[974,975] To obtain the highly resistive state, a pulse of sufficient (approximately 10V) intensity with a trailing edge faster than 100 μ sec must be applied to the formed device. The conductive state can be recaptured by applying a voltage in excess of the original threshold for negative resistance. Such behavior has been recently reviewed in detail,[833] and we shall not consider it further here.

VIII. CONCLUSIONS

It is obvious that we have come a long way in the past two years in understanding the structural, electrical, and optical properties of amorphous semiconductors. A basic band model has been developed which enables us to discuss the experimental data in terms of key parameters such as the mobility gap and the density of states at the Fermi energy. Much theoretical justification of this band model has been provided, on the basis of alloy analogies, although the real problem has not yet been handled in a satisfactory manner. Several firm conclusions can be reached with regard to each investigated material. Nevertheless, many conflicts have arisen which have proved difficult to resolve and a basic simplification of the field is still lacking.

We have learned that the quality of amorphous semiconductors is, if anything, more variable than that of crystalline semiconductors and that the ideal amorphous state is more difficult to approach than the perfect crystal. The complications of the possible existence of many different amorphous forms with a given composition is infinitely more severe than the problem of crystalline allotropes, since the diffraction patterns of the different amorphous structures appear to be indistinguishable. The problem of detecting small crystallites in amorphous solids has not yet been attacked, nor has the problem of devitrification during, for example, transport measurements or electron-beam observations. We have only begun to scratch the surface with regard to characterization in general. Until these difficulties are overcome, interpretation of the available data is extremely dangerous. Reproducibility is not a sufficient criterion for accuracy, since the same pitfalls are always present even in different laboratories.

We cannot predict the future, but we can extrapolate from the past. It is often remarked that the present-day state of the field of amorphous semiconductors is analogous to that of crystalline Ge and Si twenty years ago. This analogy appears to be appropriate. It was primarily the prospect of commercial applications which led to the expenditures of large quantities of research time and effort in understanding crystalline semiconductors, and the same hope is now present with regard to amorphous semiconductors. It could be cautioned that the problems to be faced are much more difficult in amorphous solids, but because of our advances in the last twenty years, we are now much more sophisticated.

What directions will research take in the next two years? Experimentally, there will probably be greater efforts in electrode-free measurements, such as optical, submillimeter, and microwave experiments on materials continuously held under high vacuum, sealed off from impurities. More four-probe measurements of resistivity are needed. Magneto-optical work has proved to be difficult and few modulation-spectroscopy experiments have yet been carried out. Theoretically, modern many-body techniques have not yet been introduced and electronic correlations have generally been ignored, despite their obvious importance. In any event, thoughts about the future of the field do not suggest any feeling of ennui, and these days we must be appreciative of little favors.

Acknowledgments

I have benefitted from many discussions with my colleagues on the subject of amorphous semiconductors. In particular, I should like to thank Floyd Arntz, Arthur Bienenstock, Marc Brodsky, Morrel Cohen, John deNeufville, Edward Fagen, Julius Feinleib, Helmut Fritzsche, Theodore Kaplan, Joseph Mogab, Jan Morgan, Simon Moss, Nevill Mott, Ronald Neale, Sally Nutter, William Paul, Stanford Ovshinsky, Howard Rockstad, Jean Sauvage, Robert Shaw, James Thompson, Kurt Weiser, and Richard Zallen. The original work discussed in this review was partially supported by the Advanced Research Projects Agency of the Department of Defense, and was monitored by the U.S. Army Research Office - Durham, under Contract No. DAHCO4 70 C 0048.

REFERENCES

1. Ovshinsky, S. R., *Phys. Rev. Lett.*, 21, 1450, 1968.
2. Henisch, H. K., *Sci. Amer.*, November, 1969, 30.
3. Adler, D., *Electronics*, September 28, 1970, 61.
4. In many crystalline materials, such as V_2O_3 and VO_2, a change in crystal symmetry brings about a sharp jump in electrical conductivity at the transition temperature, sometimes by a factor as large as 10^7. See, for example, Adler, D., *Rev. Mod. Phys.*, 40, 714, 1968.
5. Mott, N. F. and Twose, W. D., *Advan. Phys.*, 10, 107, 1961.
6. Mott, N. F., *Advan. Phys.*, 16, 49, 1967.
7. Borland, R. E., *Proc. Phys. Soc.*, 78, 926, 1961.
8. Borland, R. E., *Proc. Roy. Soc.*, A274, 529, 1963.
9. Borland, R. E. and Bird, N. F., *Proc. Phys. Soc.*, 83, 23, 1964.
10. Richter, H. and Furst, O., *Z. Naturforsch.*, 6a, 38, 1951.

11. Moss, S. C. and Graczyk, J. F., *Phys. Rev. Lett.*, 23, 1167, 1969.

12. It is interesting to note that this similarity does not exist between crystalline and liquid Ge. When crystalline Ge melts, short-range order persists above the melting point, but the local environment is octahedral rather than tetrahedral, similar to a simple cubic lattice. Each Ge is surrounded by six nearest neighbors, and the change in local bonding leads to liquid Ge having metallic electrical conductivity. See, for example, Krebs, H., Lazarev, V. B., and Winkler, L., *Z. Anorg. Allg. Chem.*, 352, 277, 1967.

13. Seitz, F., *J. Phys. Chem.*, 57, 737, 1953.

14. Cohen, M. H., Lectures at the NATO Summer Course on Amorphous Semiconductors, Ghent, Belgium, 1969, unpublished.

15. Born, M. and Oppenheimer, J. R., *Ann. Phys.*, 84, 457, 1927.

16. Stern, E. A., *Energy Bands in Metals and Alloys*, L. H. Bennett and J. T. Waber, Eds., Gordon and Breach, New York, 1968, 151.

17. Slater, J. C., *J. Appl. Phys.*, 39, 761, 1968.

18. DeGennes, P. G. and Friedel, J., *J. Phys. Chem. Solids*, 4, 71, 1958.

19. Born, M. and von Karman, T., *Phys. Zeit.*, 13, 297, 1912.

20. A glass is usually defined as an amorphous material that can be obtained by quenching rapidly from the liquid state, thus freezing in the structural disorder. A *chalcogenide* glass contains a high proportion of the chalcogen elements, Te, Se, and/or S.

21. Svensson, E. C., Brockhouse, B. N., and Rowe, J. M., *Phys. Rev.*, 155, 619, 1967.

22. For a review of the structure of amorphous materials, see S. C. Moss, *Solid State Phys.*, to be published.

23. Adler, D., *Solid State Phys.*, 21, 1, 1968.

24. See, for example, Blatt, F. J., *Physics of Electronic Conduction in Solids*, McGraw-Hill, New York, 1968, 117.

25. Reitz, J. R., *Solid State Phys.*, 1, 1, 1955.

26. Bloch, F., *Z. Phys.*, 52, 555, 1928.

27. Weinreich, G., *Solids*, John Wiley & Sons, New York, 1965, 33.

28. Ioffe, A. F. and Regel, A. R., *Progr. Semiconductors*, 4, 237, 1960.

29. Shockley, W., *Electrons and Holes in Semiconductors*, Van Nostrand, New York, 1950.

30. Koster, G. F. and Slater, J. C., *Phys. Rev.*, 96, 1208, 1954.

31. Mott, N. F., *Proc. Phys. Soc.*, (London) A62, 416, 1949.

32. Mott, N. F., *Phil. Mag.*, 6, 287, 1961.

33. As long as antisymmetric eigenfunctions of the form of equation (3) are used, two outer s electrons of the same spin cannot simultaneously be present on a given atom. Thus, correlations are neglected between electrons of opposite spin only.

34. Since N electrons are present, the sum of the single-particle energies is N(I + U/2), but this counts the electronic interaction energy, NU/4, twice. Subtracting the sum of the interactions, we obtain the proper total energy, N(I + U/4).

35. Adler, D. and Brooks, H., *Comments on Solid State Phys.*, 1, 145, 1968.

36. In order to construct a proper basis of localized states, we must use orthogonalized atomic orbitals. These can be constructed easily from the non-orthogonalized set. See Lowdin, P. O., *J. Chem. Phys.* 18, 365, 1950.

37. Kohn, W., *Phys. Rev.*, 133, A171, 1964.

38. Hubbard, J., *Proc. Roy. Soc.*, A276, 238, 1963.

39. Hubbard, J., *Proc. Roy. Soc.*, A277, 237, 1964.

40. Hubbard, J., *Proc. Roy. Soc.*, A281, 401, 1964.

41. Kaplan, J. A. and Argyres, P. N., *Phys. Rev.*, B1, 2457, 1970.

42. Bari, R. A., Adler, D., and Lange, R. V., *Phys. Rev.*, B2, 2898, 1970.

43. Mott, N. F., *Canad. J. Phys.*, 34, 1356, 1956.

44. Pines, D., *Solid State Phys.*, 1, 368, 1955.

45. Frohlich, H., *Advan. Phys.*, 3, 325, 1954.

46. Alcock, G. R., *Advan. Phys.*, 5, 412, 1956.

47. Appel, J., *Solid State Phys.*, 21, 193, 1968.

48. Feynman, R. P., *Phys. Rev.*, 97, 660, 1955.

49. Holstein, T., *Ann. Phys.*, (N.Y.) 8, 325, 1959.

50. Holstein, T., *Ann. Phys.*, (N.Y.) 8, 343, 1959.

51. Landau, L. D., *Phys. Z. Sowjetunion*, 3, 664, 1933.

52. Adler, D., *Proc. Tenth Int. Conf. Phys. Semiconductors, Cambridge, Mass., 1970*, U.S.A.E.C. Div. Tech. Inform., Oak Ridge, Tenn., 1970, 317.

53. Holstein, T. and Friedman, L., *Phys. Rev.*, 165, 1019, 1968.

54. For example, see Mott, N. F. and Jones, H., *Theory of the Properties of Metals and Alloys*, Dover, N.Y., 1958, 65.

55. Van Hove, L., *Phys. Rev.*, 89, 1189, 1953.

56. Lifshitz, I. M., *J. Exp. Theor. Phys.*, 12, 117, 137, 156, 1942.

57. Lifshitz, I. M., *J. Exp. Theor. Phys.*, 17, 1017, 1076, 1947.

58. Lifshitz, I. M., *J. Exp. Theor. Phys.*, 18, 293, 1948.

59. Lifshitz, I. M., *Uspekhi Mat. Nauk.*, 7, 170, 1952.

60. Lifshitz, I. M. and Stepanova, G. I., *Soviet Phys. - J.E.T.P.*, 3, 656, 1956.

61. Lifshitz, I. M. and Stepanova, G. I., *Soviet Phys. - J.E.T.P.*, 4, 150, 1957.

62. Lifshitz, I. M., *Advan. Phys.*, 13, 483, 1964.

63. Lifshitz, I. M., *Sov. Phys. - Uspekhi*, 7, 549, 1965.

64. Frohlich, H., *Proc. Roy. Soc.*, A188, 521, 1947.

65. Baltensberger, W., *Phil. Mag.*, 44, 1355, 1953.

66. Landauer, R. and Helland, J. C., *J. Chem. Phys.*, 32, 1655, 1954.

67. Stern, F. and Talley, R. M., *Phys. Rev.*, 100, 1638, 1955.

68. Parmenter, R. H., *Phys. Rev.*, 104, 22, 1956.

69. Lax, M. and Phillips, J.C., *Phys. Rev.*, 110, 41, 1958.

70. Edwards, S. F., *Phil. Mag.*, 6, 617, 1961.

71. Bonch-Bruevich, V. L. and Mironov, A. G., *Sov. Phys. - Solid State,* 3, 2194, 1962.

72. Wolff, P. A., *Phys. Rev.*, 126, 405, 1962.

73. Kane, E. O., *Phys. Rev.*, 131, 1532, 1963.

74. Gubanov, A . I., *Quantum Electron Theory of Amorphous Conductors*, Consultants Bureau, N. Y., 1965.

75. Halperin, B. I. and Lax, M., *Phys. Rev.*, 148, 722, 1966.

76. Zittarz, J. and Langer, J. S., *Phys. Rev.*, 148, 741, 1966.

77. Mott, N. F., *Phil. Mag.*, 13, 989, 1966.

78. Halperin, B. I. and Lax, M., *Phys. Rev.*, 153, 802, 1967.

79. Andreev, I. V., *Sov. Phys. - J.E.T.P.*, 21, 961, 1965.

80. Bonch-Bruevich, V. L., *J. Non-Crystalline Solids*, 4, 410, 1970.

81. Lax, M., *Rev. Mod. Phys.*, 23, 287, 1951.

82. Soven, P., *Phys. Rev.*, 156, 809, 1967.

83. Butler, W. H., Thesis, U. of California, San Diego, 1969, unpublished.

84. Butler, W. H. and Kohn, W., *J. Res. N.B.S.*, 74A, 443, 1970.

85. Freed, K. F. and Cohen, M. H., to be published.

86. Thermally activated hopping conduction can still give a non-vanishing mobility within the band tails.

87. Anderson, P. W., *Phys. Rev.*, 109, 1492, 1958.

88. Watson, K. M., *Phys. Rev.*, 105, 1388, 1957.

89. Bonch-Bruevich, V. L., *Phys. Lett.*, 18, 260, 1965.

90. Ziman, J. M., *J. Phys.*, C 2, 1230, 1969.

91. Lloyd, P., *J. Phys.*, C 2, 1717, 1969.

92. Brouers, F., *J. Non-Crystalline Solids*, 4, 428, 1970.

93. Ziman, J. M., *J. Non-Crystalline Solids*, 4, 426, 1970.

94. Anderson, P. W., *J. Non-Crystalline Solids*, 4, 433, 1970.

95. Anderson, P. W., *Comments on Solid State Phys.*, 2, 193, 1970.

96. Thouless, D. J., *J. Phys.*, C 3, 1559, 1970.

97. Mott, N. F., *Phil. Mag.*, 17, 1259, 1968.

98. Mott, N. F. and Davis, E. A., *Phil. Mag.*, 17, 1269, 1968.

99. Mott, N. F., *Rev. Med. Phys.*, 40, 677, 1968.

100. Mott, N. F., *Phil. Mag.*, 19, 835, 1969.

101. Mott, N. F., *Contemp. Phys.*, 10, 125, 1969.

102. Mott, N. F., *Festkorperprobleme*, 9, 22, 1969.

103. Mott, N. F., *Phil. Mag.*, 22, 1, 1970.

104. Edwards, S. F., *J. Non-Crystalline Solids*, 4, 417, 1970.

105. Cohen, M. H., *J. Non-Crystalline Solids*, 4, 391, 1970.

106. Williams, F. W. and Matthews, N. F. J., *Phys. Rev.*, 180, 864, 1969.

107. Economou, E. N. and Cohen, M. H., *Phys. Rev. Lett.*, 24, 218, 1970.

108. Cohen, M. H., Fritzsche, H., and Ovshinsky, S. R., *Phys. Rev. Lett.*, 22, 1065, 1969.

109. Economou, E. N. and Cohen, M. H., to be published.

110. Economou, E. N., Kirkpatrick, S., Cohen, M. H., and Eggarter, T. P., *Phys. Rev. Lett.*, 25, 520, 1970.

111. Economou, E. N. and Cohen, M. H., *Phys. Rev. Lett.*, 25, 1445, 1970.

112. Neustadter, H. E. and Coopersmith, M. H., *Phys. Rev. Lett.*, 23, 585, 1969.

113. Coopersmith, M. H., *Phys. Rev.*, 139, A1359, 1965.

114. Eggarter, T. P. and Cohen, M. H., *Phys. Rev. Lett.*, 25, 807, 1970.

115. Levine, J. L. and Sanders, T. M., *Phys. Rev.*, 154, 138, 1967.

116. Kubo, R., *J. Phys. Soc. Jap.*, 12, 570, 1957.

117. Greenwood, D. A., *Proc. Phys. Soc.*, (London), 71, 585, 1958.

118. Kronig, R. deL. and Penney, W. G., *Proc. Roy. Soc.*, A130, 499, 1930.

119. Srinivasan, G. and Cohen, M. H., *J. Non-Crystalline Solids*, 3, 393, 1970.

120. Boer, K. W., *J. Non-Crystalline Solids*, 2, 444, 1970.

121. Boer, K. W. and Haislip, R., *Phys. Rev. Lett.*, 24, 230, 1970.

122. Adler, D. and Feinleib, J., *Phys. Rev.*, B2, 3112, 1970.

123. Adler, D. and Feinleib, J., *Electronic Density of States*, N.B.S. Special Publication 323, in press.

124. Hindley, N. K., *J. Non-Crystalline Solids*, 5, 17, 1970.

125. Cohen, M. H., *J. Non-Crystalline Solids*, 2, 432, 1970.

126. Mott, N. F., *J. Non-Crystalline Solids*, 1, 1, 1968.

127. Cutler, M. and Mott, N. F., *Phys. Rev.*, 181, 1369, 1969.

128. Pollak, M. and Geballe, T. H., *Phys. Rev.*, 133, A1742, 1964.

129. Austin, I. G. and Mott, N. F., *Advan. Phys.*, 18, 41, 1969.

130. Davis, E. A. and Mott, N. F., *Phil. Mag.*, 22, 903, 1970.

131. Konig, H., *Reichsber. Phys.*, 1, 4, 1944.

132. Krikorian, E. and Sneed, R. J., *J. Appl. Phys.*, 37, 3665, 1966.

133. Szekely, G., *J. Electrochem. Soc.*, 98, 318, 1951.

134. Chittuck, R. C., Alexander, J. H., and Sterling, H. F., *J. Electrochem. Soc.*, 116, 77, 1969.

135. Richter, H. and Breitling, G., *Z. Naturforsch.*, 13a, 988, 1958.

136. Brodsky, M. H., Title, R. S., Weiser, K., and Pettit, G. D., *Phys. Rev.*, B1, 2632, 1970.

137. See Ehrenreich, H. and Turnbull, D., *Comments on Solid State Phys.*, 3, 75, 1970, for a brief discussion of these models.

138. Light, T. B. and Wagner, C. N. J., *J. Appl. Cryst.*, 1, 199, 1968.

139. Grigorovici, R., *Mat. Res. Bull.*, 3, 13, 1968.

140. Grigorovici, R. and Manaila, R., *Thin Solid Films*, 1, 343, 1968.

141. Grigorovici, R. and Manaila, R., *J. Non-Crystalline Solids*, 1, 371, 1969.

142. Grigorovici, R., Manaila, R., and Varpolin, A. A., *Acta Crystallogr.*, B24, 535, 1968.

143. Coleman, M. V. and Thomas, D. J. D., *Phys. Status Solidi*, 24, K111, 1967.

144. Polk, D. E., *J. Non-Crystalline Solids*, 5, 365, 1971.

145. Zachariasen, W. H., *J. Amer. Chem. Soc.*, 54, 3841, 1932.

146. Breitling, G., *J. Vac. Sci. Technol.*, 6, 628, 1969.

147. Clark, A. H., *Phys. Rev.*, 154, 750, 1967.

148. Light. T. B., *Phys. Rev. Lett.*, 22, 999, 1969.

149. Donovan, T. M., Spicer, W. E., and Bennett, J. M., *Phys. Rev. Lett.*, 22, 1058, 1969.

150. Donovan, T. M., Spicer, W. E., Bennett, J. M., and Ashley, E. J., *Phys. Rev.*, B2, 397, 1970.

151. Mogab, C. J. and Block, R. G., to be published.

152. Moss, S. C. and Graczyk, J. F., *Phys. Rev. Lett.*, 23, 1167, 1969.

153. Chopra, K. L., *Thin Film Phenomena*, McGraw-Hill, New York, 1969, Chap. 4.

154. Moss, S. C. and Graczyk, J. F., *Proc. Tenth Int. Conf. Phys. Semicond., Cambridge, Mass., 1970,* U.S.A.E.C. Div. Tech. Inform., Oak Ridge, Tenn., 1970, 658.

155. Cohen, M. H. and Turnbull, D., *Nature*, 203, 964, 1964.

156. Cohen, M. H., *Proc. Tenth Int. Conf. Phys. Semicond., Cambridge, Mass., 1970,* U.S.A.E.C. Div. Tech. Inform., Oak Ridge, Tenn., 1970, 645.

157. Brodsky, M. H. and Title, R. S., *Phys. Rev. Lett.*, 23, 581, 1969.

158. Crowder, B. L., Title, R. S., Brodsky, M. H., and Pettit, G. D., *Appl. Phys. Lett.*, 16, 205, 1970.

159. Title, R. S., Brodsky, M. H., and Crowder, B. L., *Proc. Tenth Int. Conf. Phys. Semicond., Cambridge, Mass., 1970*, U.S.A.E.C. Div. Tech. Inform., Oak Ridge, Tenn., 1970, 794.

160. Chung, M. G. and Haneman, D., *J. Appl. Phys.*, 37, 1879, 1966.

161. Haneman, D., *Phys. Rev.*, 170, 705, 1968.

162. Haneman, D., Chung, M. F., and Taloni, A., *Phys. Rev.*, 170, 719, 1968.

163. Brodsky, M. H., *J. Vac. Sci. Tech.*, 8, 125, 1971.

164. Hass, G. and Scott, N. W., *J. Phys. Rad.*, 11, 394, 1950.

165. Dunoyer, J. M., *J. Phys. Rad.*, 12, 602, 1951.

166. Reimer, L., *Z. Naturforsch.*, 13a, 536, 1958.

167. Richter, H. and Schneider, R., *Z. Angew. Phys.*, 11, 277, 1959.

168. Konorov, P. P. and Romanov, O. V., *Sov. Phys. - Solid State*, 2, 1688, 1960.

169. Rogowski, F., *Z. Electrochem.*, 64, 305, 1960.

170. Suhramm, R., Kruel, M., and Wedler, G., *Z. Phys.*, 173, 71, 1963.

171. Adirovich, E. I. and Yuabov, Yu. M., *Sov. Phys. - Doklady*, 9, 296, 1964.

172. Croitoru, N. and Marinescu, N., *Rev. Phys.*, (Bucharest) 9, 202, 1964.

173. Grigorovici, R., Croitoru, N., Devenyi, A., and Teleman, E., *Proc. Int. Conf. Phys. Semicond., Paris, 1964*, Dunod, Paris, 1964, 423.

174. Grigorovici, R., Croitoru, N., Devenyi, A., Vescan, L., and Barna, P., *Rev. Rom. Phys.*, 10, 649, 1965.

175. Grigorovici, R., Croitoru, N., and Devenyi, A., *Phys. Status Solidi*, 23, 621, 1967.

176. Walley, P. A. and Jonscher, A. K., *Thin Solid Films*, 1, 367, 1967.

177. Grigorovici, R. and Vancu, A., *Thin Solid Films*, 2, 105, 1968.

178. Walley, P. A., *Thin Solid Films*, 2, 327, 1968.

179. Jonscher, A. K. and Walley, P. A., *J. Vac. Sci. Tech.*, 6, 662, 1969.

180. Stuke, J., *Proc. Conf. on Electronic Processes in Low-Mobility Solids, Sheffield, 1966*, U. of Sheffield Dept. of Glass Technology, Sheffield, 1966, 77.

181. Stuke, J., *J. Non-Crystalline Solids*, 4, 1, 1970.

182. Stuke, J., *Proc. Tenth Int. Conf. Phys. Semicond., Cambridge, Mass., 1970*, U.S.A.E.C. Div. Tech. Inform., Oak Ridge, Tenn., 1970, 14.

183. Chopra, K. L. and Bahl, S. K., *Phys. Rev.*, B1, 2545, 1970.

184. Rusu, C., Devenyi, A., and Rusu, M., *Proc. Cong. Int. Couches Minces, Cannes, 1970*, S.F.I.T.V., Paris, 1970, p. 501.

185. Piller, H. and Khan, S. A., *Proc. Tenth Int. Conf. Phys. Semicond., Cambridge, Mass., 1970,* U.S.A.E.C. Div. Tech. Inform., Oak Ridge, Tenn., 1970, 662.

186. Beyer, W., Mell, H., and Stuke, J., to be published.

187. Nwachuku, A. and Kuhn, M., *Appl. Phys. Lett.*, 12, 163, 1968.

188. Grigorovici, R., Croitoru, N., and Devenyi, A., *Phys. Status Solidi*, 16, K143, 1966.

189. Grigorovici, R., Croitoru, N., and Devenyi, A., *Rev. Rom. Phys.*, 11, 869, 1966.

190. Spear, W. E., *J. Non-Crystalline Solids*, 1, 197, 1969.

191. Le Comber, P. G. and Spear, W. E., *Phys. Rev. Lett.*, 25, 509, 1970.

192. Dolezalek, F. K. and Spear, W. E., *J. Non-Crystalline Solids*, 4, 97, 1970.

193. However, note that most of these measurements were performed on amorphous Ge rather than Si.

194. Osmun, J. W. and Fritzsche, H., *Appl. Phys. Lett.*, 16, 87, 1970.

195. Sauvage, J., Mogab, C. J., and Adler, D., to be published.

196. Bermon, S., personal communication.

197. Ovsyuk, V. N. and Rzhanov, A. V., *Sov. Phys. - Semicond.*, 3, 250, 1966.

198. Sauvage, J., Mogab, C. J., and Adler, D., *J. Non-Crystalline Solids*, in press.

199. Kastner, M. and Fritzsche, H., *Mat. Res. Bull.*, 5, 631, 1970.

200. Many, A., Goldstein, Y., and Grover, N. B., *Semiconductor Surfaces*, North-Holland, Amsterdam, 1965.

201. Grigorovici, R. and Devenyi, A., *Proc. Ninth Int. Conf. Phys. Semicond., Moscow, 1968*, Vol. 2, Nauka, Leningrad, 1968, 1267.

202. Devenyi, A., Belu, A., and Korony, G., *J. Non-Crystalline Solids*, 4, 380, 1970.

203. Grigorovici, R., Devenyi, A., and Belu, A., *Proc. Congres. Int. Couches Minces, Cannes, 1970*, S.F.I.T.V., Paris, 1970, 463.

204. Fuhs, W. and Stuke, J., *Mat. Res. Bull.*, 5, 611, 1970.

205. Keyes, R. W., *Solid State Phys.*, 11, 149, 1960.

206. Smith, C. S., *Phys. Rev.*, 94, 42, 1954.

207. Devenyi, A., Belu, A., and Sladaru, S., *Thin Solid Films*, 4, 211, 1969.

208. Morin, F. J., Geballe, T. H., and Herring, C., *Phys. Rev.*, 105, 525, 1957.

209. Paul, W., *J. Phys. Chem. Solids*, 8, 196, 1959.

210. Mell, H. and Stuke, J., *J. Non-Crystalline Solids*, 4, 304, 1970.

211. Fritzsche, H. and Lark-Horovitz, K., *Phys. Rev.*, 99, 400, 1955.

212. Sasaki, W., *J. Phys. Soc. Jap.*, 20, 825, 1965.

213. Yamanouchi, C., Mizuguchi, K., and Sasaki, W., *J. Phys. Soc. Jap.*, 22, 859, 1967.

214. Toyozawa, Y., *J. Phys. Soc. Jap.*, 17, 986, 1962.

215. Toyozawa, Y., *Proc. Int. Conf. Phys. Semicond., Exeter, 1962*, The Institute of Physics and the Physical Society, London, 1962, 104.

216. Alexander, M. N. and Holcomb, D. F., *Rev. Mod. Phys.*, 40, 815, 1968.

217. Bosnell, J. R. and Voisey, U. C., *Thin Solid Films*, 6, 161, 1970.

218. Mort, J., *Phys. Rev.*, B3, 3576, 1971.

219. Shockley, W., *Electrons and Holes in Semiconductors*, Van Nostrand, New York, 1950, 211.

220. Tauc, J., Abraham, A., Pajasova, L., Grigorovici, R., and Vancu, A., *Proc. Conf. Physic of Non-crystalline Solids, Delft, 1964*, Ed., J. A. Prins, North-Holland, Amsterdam, 1965, 606.

221. Glass, A. M., *Canad. J. Phys.*, 43, 1068, 1965.

222. Tauc, J., Grigorovici, R., and Vancu, A., *Phys. Status Solidi*, 15, 627, 1966.

223. Tauc, J., Grigorovici, R., and Vancu, A., *J. Phys. Soc. Jap.*, 21, Suppl., 123, 1966.

224. Wales, J., Lowitt, G. J., and Hill, R. A., *Thin Solid Films*, 1, 137, 1967.

225. Tauc, J., *Mat. Res. Bull.*, 3, 37, 1968.

226. Grigorovici, R. and Vancu, A., *Thin Solid Films*, 2, 105, 1968.

227. Tauc, J., *Optical Properties of Solids*, Abeles, F., Ed., North-Holland, Amsterdam, 1969, 123.

228. Spicer, W. E. and Donovan, T. M., *J. Non-Crystalline Solids*, 2, 66, 1970.

229. Beaglehole, D. and Zavetova, M., *J. Non-Crystalline Solids*, 4, 272, 1970.

230. Theye, M. L., *Mat. Res. Bull.*, 6, 103, 1971.

231. Connell, G. A. N. and Paul, W., *Bull. Amer. Phys. Soc.*, 16, 347, 1971.

232. Paul, W., Connell, G. A. N., and Shevchik, N. J., *Bull. Amer. Phys. Soc.*, 16, 347, 1971.

233. Herman, F., Kortum, R. L., and Kuglin, C. D., *Int. J. Quant. Chem.*, 15, 533, 1967.

234. Philipp, H. R. and Ehrenreich, H., *Phys. Rev.*, 129, 1550, 1963.

235. Cohen, M. L. and Bergstresser, T. K., *Phys. Rev.*, 141, 789, 1966.

236. Tauc, J., Abraham, A., Zallen, R., and Slade, M., *J. Non-Crystalline Solids*, 4, 279, 1970.

237. Dash, W. E. and Newman, R., *Phys. Rev.*, 99, 1151, 1955.

238. Hobden, M. V., *J. Phys. Chem. Solids*, 23, 821, 1962.

239. Paul, W., *J. Appl. Phys.*, 32, 2082, 1961.

240. Melz, P. J., to be published.

241. Adler, D., to be published.

242. Fan, H. Y. and Lark-Horovitz, K., *Rep. Bristol Conf. on Defects in Crystalline Solids, Physical Society,* London, 1955, 232.

243. Cleland, T. W., Crawford, J. H., and Holmes, D. K., *Phys. Rev.*, 102, 722, 1956.

244. James, H. M. and Lark-Horovitz, K., *Z. Phys. Chem.*, 198, 107, 1951.

245. Crawford, J. H. and Cleland, T. W., *Progr. Semiconductors*, 2, 67, 1957.

246. Longo, T. A. and Kleitman, D., Purdue University Progress Reports to Signal Corps P.R.F. 1046, April, 1955, unpublished.

247. Donovan, T. M. and Spicer, W. E., *Phys. Rev. Lett.*, 21, 1572, 1968.

248. Peterson, C. W., Dinan, J. H., and Fischer, T. E., *Phys. Rev. Lett.*, 25, 861, 1970.

249. Shay, J. L. and Spicer, W. E., *Phys. Rev.*, 161, 799, 1967.

250. Piller, H., Seraphin, B. O., Markel, K., and Fischer, J. E., *Phys. Rev. Lett.*, 23, 775, 1969.

251. Cardona, M., *J. Appl. Phys.*, 32, 2151, 1961.

252. Lippincott, E. R., Van Valkenburg, A., Weir, C. E., and Bunting, E. N., *J. Res. Nat. Bur. Stand.*, 61, 61, 1958.

253. Whan, R. W. and Stein, H. J., *Appl. Phys. Lett.*, 3, 187, 1963.

254. Brockhouse, B. N. and Iyengar, P. K., *Phys. Rev.*, 111, 747, 1958.

255. Zallen, R., *Phys. Rev.*, 173, 824, 1968.

256. Smith, S. D. and Hardy, J. R., *Phil. Mag.*, 5, 1311, 1960.

257. Fan, H. Y. and Ramadas, A. K., *J. Appl. Phys.*, 30, 1127, 1959.

258. Smith, J. E., Jr., Brodsky, M. H., Crowder, B. L., Nathan, M. I., and Pinczuk, A., *Phys. Rev. Lett.*, 26, 642, 1971.

259. Shuker, R. and Gammon, R. W., *Phys. Rev. Lett.*, 25, 222, 1970.

260. Dolling, G. and Cowley, R. A., *Proc. Phys. Soc.*, (London) 88, 463, 1966.

261. Parker, J. H., Jr., Feldman, D. W., and Ashkin, M., *Phys. Rev.*, 155, 712, 1967.

262. Beserman, R., Sebenne, C., and Balkanski, M., *Proc. Congres Int. Couches Minces, Cannes, 1970*, S.F.I.T.V., Paris, 1970, 509.

263. Herman, F. and Van Dyke, J. P., *Phys. Rev. Lett.*, 21, 1575, 1968.

264. Herman, F. and Van Dyke, J. P., *J. Non-Crystalline Solids*, 2, 81, 1970.

265. Herman, F., Kortum, R. L., Kuglin, C. D., and Short, R. A., *Quantum Theory of Atoms, Molecules, and the Solid State*, Lowdin, P. O., Ed., Academic Press, New York, 1966, 381.

266. Brust, D., *Phys. Rev.*, 186, 768, 1969.

267. Brust, D., *Phys. Rev. Lett.*, 23, 1232, 1969.

268. Spicer, W. E. and Donovan, T. M., *Phys. Rev. Lett.*, 24, 595, 1970.

269. Spicer, W. E. and Donovan, T. M., *Proc. Tenth Int. Conf. Phys. Semicond., Cambridge, Mass., 1970.*, U.S.A.E.C. Div. Tech. Inform., Oak Ridge, Tenn., 1970, 677.

270. Chakraverty, B. K., *Proc. Congres Int. sur les Couches Minces, Cannes, October, 1970*, S.F.I.T.V., Paris, 1970, 427.

271. Leman, G. and Friedel, J., *J. Appl. Phys., Suppl.*, 33, 281, 1962.

272. Stern, F., *Phys. Rev.*, B3, 2636, 1971.

273. Kane, E. O., *Phys. Rev.*, 131, 79, 1963.

274. Cardona, M. and Pollack, F. H., *Phys. Rev.*, 142, 530, 1966.

275. Fritzsche, H., *J. Non-Crystalline Solids*, 6, 49, 1971.

276. See also Keldysh, L. V., and Proshko, G. P., *Soviet Phys. - Solid State*, 5, 2481, 1964.

277. Zallen, R. and Scher, H., to be published.

278. Clark, A. H., *J. Non-Crystalline Solids*, 2, 52, 1970.

279. Oki, F., Ogaka, Y., and Fujiki, Y., *Jap. J. Appl. Phys.*, 8, 1056, 1969.

280. Brodsky, M. H. and Turnbull, D., *Bull. Amer. Phys. Soc.*, 16, 304, 1971.

281. Lucovsky, G. and Tabak, M. D., *Selenium,* Cooper, W. C. and Zingaro, R. A., Eds., Van-Nostrand-Reinhold, New York, 1971, in press.

282. Dessauer, J. H. and Clark, H. E., *Xerography and Related Processes*, Focal Press, London, 1965.

283. Schaffert, R. M., *Electrophotography*, Focal Press, London, 1965.

284. Marsh, R. E., Pauling, L., and McCullough, J. D., *Acta Crystallogr.*, 6, 71, 1953.

285. Burbank, R. D., *Acta Crystallogr.*, 4, 140, 1951.

286. Burbank, R. D., *Acta Crystallogr.*, 5, 236, 1952.

287. Briegleb, G., *Z. Phys. Chem.*, A144, 321, 1929.

288. Krebs, H. and Morsch, W., *Z. Anorg. Allg. Chem.*, 263, 305, 1950.

289. Krebs, H., *Z. Anorg. Allg. Chem.*, 265, 156, 1951.

290. Gee, G., *Trans. Faraday Soc.*, 48, 515, 1952.

291. Richter, H., Kulcke, W., and Specht, H., *Z. Naturforsch.*, 7a, 511, 1952.

292. Eisenberg, A. and Tobolsky, A. V., *J. Polym. Sci.*, 56, 19, 1960.

293. Eisenberg. A. and Tobolsky, A. V., *J. Polym. Sci.*, 61, 483, 1962.

294. Srb, I. and Vasko, A., *Czech. J. Phys.*, B 13, 827, 1963.

295. Hendus, H., *Z. Phys.*, 119, 265, 1942.

296. Krebs, H. and Schultze-Gebhardt, F., *Acta Crystallogr.*, 8, 412, 1955.

297. Grimminger, H., Gruninger, H., and Richter, H., *Naturwissenschaften*, 42, 256, 1955.

298. Andrievskii, A. I., Nabitovitch, I. D., and Voloshchuk, Ya. V., *Sov. Phys. - Cryst.*, 5, 349, 1960.

299. Henninger, E. H., Buschert, R. C., and Heaton, L., *J. Chem. Phys.*, 46, 586, 1967.

300. Kaplow, R., Rowe, T. A., and Averbach, B. L., *Phys. Rev.*, 168, 1068, 1968.

301. Breitling, G. and Richter, H., *Mat. Res. Bull.*, 4, 19, 1969.

302. Tobolsky, A., Owen, G., and Eisenberg, A., *J. Colloid Sci.*, 17, 717, 1962.

303. Eisenberg, A. and Teter, L., *J. Amer. Chem. Soc.*, 87, 2108, 1965.

304. Myers, M. B. and Felty, E. J., *Mat. Res. Bull.*, 2, 715 misprinted as 535, 1967.

305. Tobolsky, A. and Murakami, K., *J. Polym. Sci.*, 40, 443, 1959.

306. Massen, C. H., Weijts, A. G. L. M., and Poulis, J. A., *Trans. Faraday Soc.*, 60, 317, 1964.

307. Caldwell, R. S. and Fan, H. Y., *Phys. Rev.*, 114, 664, 1959.

308. Siemsen, K. J., *J. Phys. Chem. Solids*, 30, 1897, 1969.

309. Lucovsky, G., Mooradian, A., Taylor, W., Wright, G. B., and Keezer, R. C., *Solid State Commun.*, 5, 113, 1967.

310. Lucovsky, G., *Physics of Selenium and Tellurium*, Cooper, W. C., Ed., Pergamon Press, Oxford, 1969, 255.

311. Lucovsky, G., *Mat. Res. Bull.*, 4, 505, 1969.

312. Zallen, R. and Lucovsky, G., *Selenium*, Cooper, W. C. and Zingaro, R. A., Eds., Van Nostrand-Reinhold, New York, 1971, in press.

313. Mooradian, A. and Wright, G. B., *Physics of Selenium and Tellurium*, Cooper, W. C., Ed., Pergamon Press, Oxford, 1969, 269.

314. Keezer, R. C. and Bailey, M. W., *Mat. Res. Bull.*, 2, 185, 1967.

315. Shirai, T., Hamada, S., and Kobayaski, K., *Nippon Kagako Zasshi,* 84, 968, 1963.

316. Schottmiller, J., Tabak, M., Lucovsky, G., and Ward, A., *J. Non-Crystalline Solids*, 4, 80, 1970.

317. Tobolsky, A. V. and Owen, G. D. T., *J. Polym. Sci.*, 59, 329, 1962.

318. Borelius, G., Pihlstrand, F., Anderson, L., and Gullberg, K., *Ar. Mat. Astr. Och. Fysik.*, 30A, 1, 1944.

319. Nasledov, D. N. and Malyshev, E. K., *J. Tech. Phys.*, (Moscow) 16, 1127, 1946.

320. Henkels, H. W., *J. Appl. Phys.*, 21, 725, 1950.

321. Eckhart, F., *Ann. Phys.*, (Leipzig) 14, 233, 1954.

322. Konozenko, I. D., *Advan. Phys. Sci.*, (Moscow) 52, 561, 1954.

323. Hartke, J. L., *Phys. Rev.*, 125, 1177, 1962.

324. Lanyon, H.P.D., *Phys. Rev.*, 130, 134, 1963.

325. Rossiter, E. L. and Warfield, G., Device Phys. Lab. Tech. Rep. No. 10, Dept. of Electrical Engineering, Princeton U., 1967, unpublished.

326. Lakatos, A. I. and Abkowitz, M., *Phys. Rev.*, B3, 1791, 1971.

327. Lacourse, W. C., Twaddell, V. A., and Mackenzie, J. D., *J. Non-Crystalline Solids*, 3, 234, 1970.

328. Kolomiets, B. T., Mazarova, T. F., and Shylo, V. P., *Proc. Int. Conf. Phys. Semicond., Exeter, 1962.*, The Institute of Physics and the Physical Society, London, 1962, 159.

329. Lizell, B., *J. Chem. Phys.*, 20, 672, 1952.

330. Drews, R. E., Zallen, R., and Keezer, R. C., *Bull. Amer. Phys. Soc.*, 13, 454, 1968.

331. Edmond, J. T., *Brit. J. Appl. Phys.*, 17, 979, 1966.

332. Rose, A., *Phys. Rev.*, 97, 1538, 1955.

333. Gobrecht, H., Tausend, A., and Reimers, P., Proc. Spring Meeting German Physical Soc., Freudenstadt, 1965, unpublished.

334. Stuke, J., *Festkorperprobleme*, 9, 46, 1969.

335. Dutchak, Ya. I., Kluyres, I. D., and Prokhorenko, V. Ya., *Phys. Status, Solidi*, 31, 25, 1969.

336. Borelius, G., Pihlstrand, F., Anderson, J., and Gullberg, K., *Arkiv. Mat. Astron. Fysik*, 30A, 14, 1944.

337. Henkels, H. W. and Maczuk, J., *J. Appl. Phys.*, 24, 1056, 1953.

338. Abdullaev, G. B., Aliev, G. M., Barkinkhoev, Kh. G., Askerov, Ch. M., and Larionkina, L. S., *Sov. Phys. - Solid State*, 6, 786, 1964.

339. Abdullaev, G. B., Mekhtieva, S. I., Abdinov, D. Sh., and Aliev, G. M., *Phys. Status Solidi*, 11, 891, 1965.

340. Spear, W. E., *Proc. Phys. Soc.*, London B70, 1139, 1957.

341. Spear, W. E., *Proc. Phys. Soc.*, London B76, 826, 1960.

342. Spear, W. E. and Lanyon, H. P. D., *Proc. Int. Conf. Phys. Semicond., Prague, 1960*, Czechoslovakian Academy of Sciences, Prague, 1961, 987.

343. Tabak, M. D. and Warter, P. J., *Phys. Rev.*, 173, 899, 1968.

344. Grunwald, H. P. and Blakney, R. M., *Phys. Rev.*, 165, 1006, 1968.

345. Tabak, M. D., Proc. *Third Int. Conf. Photoconductivity, Palo Alto, Calif., 1969*, Pergamon, Oxford, 1970, 87.

346. Scharfe, M. E. and Tabak, M. D., *J. Appl. Phys.*, 40, 3230, 1969.

347. Schottmiller, J., Tabak, M. D., Lucovsky, G., and Ward, A., *J. Non-Crystalline Solids*, 4, 80, 1970.

348. Tabak, M. D., *Phys. Rev.*, B2, 2104, 1970.

349. Dolezalek, F. K. and Spear, W. E., *J. Non-Crystalline Solids*, 4, 97, 1970.

350. Mort, J. and Lakatos, A. I., *J. Non-Crystalline Solids*, 4, 117, 1970.

351. Pai, D. M., *J. Chem. Phys.*, 52, 2285, 1970.

352. Mort. J., *Phys. Rev. Lett.*, 18, 540, 1967.

353. Kozyrev, P. T., *Sov. Phys. - Solid State*, 1, 94, 1959.

354. Krischunas, V. Yu. and Daukantaite, O. K., *Sov. Phys. - Solid State*, 8, 471, 1966.

355. Jamin, J., *Ann. Chim. Phys.*, 29, 303, 1850.

356. Wood, R. W., *Phil. Mag.*, 3, 607, 1902.

357. Edmunds, C. K., *Phys. Rev.*, 18, 193, 1904.

358. Meier, W., *Ann. Phys.*, 31, 1017, 1910.

359. Merwin, H. E. and Larsen, E. S., *Amer. J. Sci.*, 34, 42, 1912.

360. Foersterling, K. and Foersterling, V., *Ann. Phys.* 43, 1227, 1914.

361. Monch, G., *Phys. Z.*, 40, 487, 1939.

362. Becker, A. and Schaper, I., *Z. Phys.*, 122, 49, 1944.

363. Soezima, Y., *J. Sci. Res. Inst.*, (Tokyo) 44, 68, 1949.

364. Gebbie, H. A. and Saker, E. W., *Proc. Phys. Soc.*, (London) B64, 360, 1951.

365. Dowd, J. J., *Proc. Phys. Soc.*, (London) B64, 783, 1951.

366. Gilleo, M. A., *J. Chem. Phys.*, 19, 1291, 1951.

367. Saker, E. W., *Proc. Phys. Soc.*, (London) B65, 785, 1952.

368. Gebbie, H. A. and Cannon, C. G., *J. Opt. Soc. Amer.*, 42, 277, 1952.

369. Stuke, J., *Z. Phys.*, 134, 194, 1953.

370. Frerichs, R., *J. Opt. Soc. Amer.*, 43, 1153, 1953.

371. Hilsum, C., *Proc. Phys. Soc.*, (London) B69, 506, 1956.

372. Fochs, P. D., *Proc. Phys. Soc.*, (London) B69, 70, 1956.

373. Koehler, W. F., Odencrantz, F. K., and White, W. C., *J. Opt. Soc. Amer.*, 49, 109, 1959.

374. Gobrecht, H. and Tausend, A., *Z. Phys.*, 161, 205, 1961.

375. Kamprath, W., *Ann. Phys.*, 9, 382, 1962.

376. Hartke, J. L. and Regensburger, P. J., *Phys. Rev.*, 139, A970, 1965.

377. Siemsen, K. J. and Fenton, E. W., *Phys. Rev.*, 161, 632, 1967.

378. Kandare, S., *C. R. Acad, Sci.*, (Paris) 244, 571, 1957.

379. Weimer, P. K., *Phys. Rev.*, 79, 171, 1950.

380. Weimer, P. K. and Cope, A. D., *R. C. A. Rev.*, 12, 314, 1951.

381. Moss, T. S., *Photoconductivity in the Elements*, Butterworths, London, 1952, 187.

382. Keck, P. H., *J. Opt. Soc. Amer.*, 42, 221, 1952.

383. Moss, T. S., *Optical Properties of Semiconductors*, Butterworths, London, 1959, 160.

384. Forthand, R. A., *J. Appl. Phys.*, 31, 1558, 1960.

385. Dresner, *J. Chem. Phys.*, 35, 1628, 1961.

386. Li, H. T., and Regensburger, P. J., *J. Appl. Phys.*, 34, 1730, 1963.

387. Pai, D. and Ing, S. W., *Phys. Rev.*, 173, 729, 1968.

388. Tabak, M. D. and Scharfe, M. E., *J. Appl. Phys.*, 41, 2114, 1970.

389. Frenkel, J., *Phys. Rev.*, 54, 647, 1938.

390. Davis, E. A., *J. Non-Crystalline Solids*, 4, 107, 1970.

391. Moll, J. L., *Physics of Semiconductors*, McGraw-Hill, New York, 1964, 70.

392. Lucovsky, G., *Proc. Tenth Int. Conf. Phys. Semicond., Cambridge, Mass., 1970*, U.S.A.E.C. Div. Tech. Inform., Oak Ridge, Tenn., 1970, 799.

393. Frenkel J., *Phys. Rev.*, 37, 1276, 1931.

394. Prosser, V., *Proc. Int. Conf. Phys. Semicond., Prague, 1960*, Czechoslovakian Academy of Sciences, Prague, 1961, 993.

395. Spear, W. E. and Adams, A. R., *J. Phys. Chem. Solids*, 27, 281, 1966.

396. Cook, B. E. and Spear, W. E., *J. Phys. Chem. Solids*, 30, 1125, 1969.

397. Chen, I., *Phys. Rev.*, B2, 1053, 1970.

398. Chen, I., *Phys. Rev.*, B2, 1060, 1970.

399. Weiser, G. and Stuke, J., *Phys. Status Solidi*, 35, 747, 1969.

400. Weiser, G. and Stuke, J., *Proc. Ninth Int. Conf. Phys. Semicond.*, Moscow, 1968, Nauka, Leningrad, 1968, I, 228.

401. Stuke, J. and Weiser, G., *Phys. Status Solidi*, 17, 343, 1966.

402. Aoki, M. and Okano, S., *Jap. J. Appl. Phys.*, 5, 980, 1966.

403. Drews, R. E., *Appl. Phys. Lett.*, 9, 347, 1966.

404. Franz, W., *Z. Naturforsch*, 13, 484, 1958.

405. Keldysh, L. V., *Sov. Phys. - J.E.T.P.*, 7, 788, 1958.

406. Mort, J. and Scher, H., *Phys. Rev.*, B3, 334, 1971.

407. Becquerel. F., *Ann. Chim. Phys.*, 12, 1, 1877.

408. Gobrecht, H., Tausend, A., and Bach, I., *Z. Phys.*, 166, 76, 1962.

409. Garben, B. and Seliger, H., *Phys. Status Solidi*, 29, K27, 1968.

410. Dresner, J. and Stringfellow, G. B., *J. Phys. Chem. Solids*, 29, 303, 1968.

411. Paribok-Aleksandrovich, I. A., *Uch. Zap. VGPI*, 36, 182, 1968.

412. Paribok-Aleksandrovich, I. A., *Sov. Phys. - Solid State*, 11, 1631, 1970.

413. Kramer, B., Maschke, K., Thomas, P., and Treusch, J., *Phys. Rev. Lett.*, 25, 1020, 1970.

414. Sandrock, R., *Phys. Rev.*, 169, 642, 1968.

415. Leiga, A. G., *J. Opt. Soc. Amer.*, 58, 1441, 1968.

416. Stuke, J. and Zimmerer, G., unpublished data, quoted in ref. 181.

417. Tutihasi, S. and Chen, I., *Phys. Rev.*, 158, 623, 1967.

418. Mohler, E., Stuke, J., and Zimmerer, G., *Phys. Status Solidi*, 22, K49, 1967.

419. Phillips, J. C., *Solid State Phys.*, 18, 55, 1966.

420. Vaipolin, A. A., *Sov. Phys. - Crystallography*, 10, 509, 1966.

421. Owen, A. E., *Contemp. Phys.*, 11, 257, 1970.

422. Zallen, R., Slade, M. L. and Ward, A. T., *Phys. Rev.*, B3, 4257, 1971.

423. Pauling, L., *The Nature of the Chemical Bond*, Cornell University Press, Ithaca, N. Y., 1960.

424. Evans, B. L. and Young, P. A., *Proc. Roy. Soc.*, (London) A297, 230, 1967.

425. Vaipolin, A. A. and Porai-Koshits, E. A., *Sov. Phys. - Solid State*, 2, 1500, 1960.

426. Vaipolin, A. A. and Porai-Koshits, E. A., *Sov. Phys. - Solid State*, 5, 178, 1963.

427. Vaipolin, A. A. and Porai-Koshits, E. A., *Sov. Phys. - Solid State*, 5, 186, 1963.

428. Vaipolin, A. A. and Porai-Koshits, E. A., *Sov. Phys. - Solid State*, 5, 497, 1963.

429. Petz, J. I., Kruh, R. F., and Amstutz, G. C., *J. Chem. Phys.*, 34, 526, 1961.

430. Hopkins, T. E., Pasternak, R. A., Gould, E. S., and Herndon, J. R., *J. Phys. Chem.*, 66, 733, 1962.

431. Pozdnev, V. P. and Kolomiets, B. T., *Sov. Phys. - Solid State*, 2, 23, 1960.

432. Kolomiets, B. T., *Phys. Status Solidi*, 7, 359, 1964.

433. Tarasov, V. V., *Zh. Fiz. Khimii*, 32, 2221, 1958.

434. Shilo, V. P., and Kolomiets, B. T., *Steklo i Keram.*, 8, 10, 1963.

435. Black, J., Conwell, E. M., Sleigle, L., and Spencer, L. W., *J. Phys. Chem. Solids*, 2, 240, 1957.

436. Edmund, J. T., Anderson, A., and Gebbie, H. A., *Proc. Phys. Soc.*, 81, 378, 1963.

437. Billian, C. J. and Jerger, J., Servo Corp. of America Final Tech. Rep., Contract No. NONR 3647 (00), 1963, unpublished.

438. Felty, E. J., Lucovsky, G., and Myers, M. B., *Solid State Communications*, 5, 555, 1967.

439. Hilton, A. R., *J. Non-Crystalline Solids*, 2, 28, 1970.

440. Austin, I. G. and Garbett, E. S., *Phil. Mag.*, 23, 17, 1971.

441. Vengel, T. N. and Kolomiets, B. T., *Soviet Phys. -Tech. Phys.*, 2, 2314, 1957.

442. Uphoff, H. L. and Healy, J. H., *J. Appl. Phys.*, 32, 950, 1961.

443. Edmond, J. T., *Brit. J. Appl. Phys.*, 17, 979, 1966.

444. Owen, A. E., *Glass Ind.*, 48, 637, 1967.

445. Owen, A. E., *Glass Ind.*, 48, 695, 1967.

446. Davis, E. A. and Shaw, R. F., *J. Non-Crystalline Solids*, 2, 406, 1970.

447. Shaw, R. F., Liang, W. Y., and Yoffe, A. D., *J. Non-Crystalline Solids*, 4, 29, 1970.

448. Shaw, R. F., to be published.

449. Edmond, J. T., *J. Non-Crystalline Solids*, 1, 39, 1969.

450. Haisty, R. W. and Krebs, H., *J. Non-Crystalline Solids*, 1, 427, 1969.

451. Mamontova, T. N., Nazarova, T. F., and Kolomiets, B. T., *Vitreous State*, Izd. Akad. Nauk. S.S.S.R., 1960.

452. Kolomiets, B. T., *Phys. Status Solidi*, 7, 713, 1964.

453. Gubanov, A. I., *Sov. Phys. - Tech. Phys.*, 2, 2335, 1957.

454. Fisher, I. Z., *Sov. Phys. - Solid State*, 1, 171, 1959.

455. Kolomiets, B. T., Nazarova, T. F., and Shilo, V. P., *Proc. Int. Conf. Phys. Semicond., Exeter, 1962,* The Institute of Physics and the Physical Society, London, 1962, 259.

456. Danilov, A. V., *J. Appl. Chem. U.S.S.R.*, 37, 1120, 1965.

457. Danilov, A. V. and El Mosli, M., *Soviet Phys. - Solid State*, 5, 1472, 1964.

458. Kolomiets, B. T., Rukhlyadev, Yu. V., and Shilo, V. P., *J. Non-Crystalline Solids*, 5, 389, 1971.

459. Kolomiets, B. T., Rukhlyadev, Yu. V., and Shilo, V. P., *J. Non-Crystalline Solids*, 5, 402, 1971.

460. Bobrov, A. I., Borisova, Z. U., and Kozhina, I. I., *Vestn. Leningrad Univ.*, 10, 86, 1965.

461. Orlova, G. M., Nikandorva, G. A., Borisova, Z. U., and Epimakhova, G. N., *Inorg. Mater.*, 6, 1700, 1970.

462. Owen, A. E. and Robertson, J. M., *J. Non-Crystalline Solids*, 2, 40, 1970.

463. Johnson, V. A., *Progress in Semiconductors*, Vol. I, Gibson, A. F., Ed., Heywood, London, 1956, 65.

464. Grant, A. J. and Yoffe, A. D., *Solid State Commun.*, 8, 1919, 1970.

465. Mansfield, R., *Proc. Phys. Soc.*, (London) B69, 76, 1956.

466. Kolomiets, B. T. and Lebedev, E. A., *Sov. Phys. - Semiconductors*, 1, 244, 1967.

467. Kruglov, V. I., Strachov, I. P., and Grisin, N. A., *Vestn. Leningrad Univ.*, 10, 63, 1968.

468. Marshall, J. M., unpublished; quoted in ref. 462.

469. Viscakas, J. K., Montrimas, E. A., and Payera, A. A., *Appl. Opt. Suppl.*, 3, 79, 1969.

470. Scharfe, M. E., *Phys. Rev.*, B2, 5025, 1970.

471. Kitao, M., Araki, F., and Yamadai, S., *Phys. Status Solidi*, 37, K119, 1970.

472. Ivkin, E. B. and Kolomiets, B. T., *J. Non-Crystalline Solids*, 3, 41, 1970.

473. Taylor, P. C., Bishop, S. G., and Mitchell, D. L., *Solid State Commun.*, 8, 1783, 1970.

474. Crevecoeur, C., and deWit, H. J., *Solid State Commun.*, 9, 445, 1971.

475. Street, R. A., Davies, G., and Yoffe, A. D., *J. Non-Crystalline Solids*, 5, 276, 1971.

476, Freeman, L. A., Shaw, R. F., and Yoffe, A. D., *Thin Solid Films*, 3, 367, 1969.

477. Pollack, M., *Phil. Mag.*, 23, 519, 1971.

478. Owen, A. E., *J. Non-Crystalline Solids*, 4, 78, 1971.

479. Cole, K. S. and Cole, R. H., *J. Chem. Phys.*, 9, 341, 1941.

480. Main, C. and Owen, A. E., *Phys. Status. Solidi*, (a) 1, 297, 1970.

481. Freeman, J. J., *Principles of Noise*, John Wiley & Sons, New York, 1958.

482. Agarwal, S. C. and Fritzsche, H., *Bull. Amer. Phys. Soc.*, 15, 244, 1970.

483. Fritzsche, H., *J. Non-Crystalline Solids*, 6, 49, 1971.

484. Tauc, J., Menth, A., and Wood, D. L., *Phys. Rev. Lett.*, 25, 749, 1970.

485. Efstathiou, A., and Levin, E. R., *J. Opt. Soc. Amer.*, 58, 373, 1968.

486. Zallen, R., Drews, R. E., Emerald, R. L., and Slade, M. L., *Phys. Rev. Lett.*, 26, 1564, 1971.

487 Felty, E. J. and Myers, M. B., unpublished; quoted in ref. 462.

488. Urbach, F., *Phys. Rev.*, 92, 1324, 1953.

489. Tauc, J., Stourac, L., Vorlicek, V., and Zavetova, M., *Proc. Ninth Int. Conf. Phys. Semicord., Moscow, 1968*, Nauka, Leningrad, 1968, 1251.

490. Hopfield, J. J., *Comments on Solid State Phys.*, 1, 16, 1968.

491. Kolomiets, B. T., Mazets, T. F., and Efendiev, Sh. M., *Sov. Phys.-Semicond.*, 4, 934, 1970.

492. Kolomiets, B. T., Mamontova, T. N., and Stepanov, G. I., *Sov. Phys.-Solid State*, 7, 1320, 1967.

493. Kolomiets, B. T. and Lyubin, V. M., *Sov. Phys.-Doklady*, 4, 1345, 1959.

494. Kolomiets, B. T., Lyubin,V. M., and Averjanov, V. L., *Mat. Res. Bull.*, 5, 655, 1970.

495. Kolomiets, B. T. and Raspopova, E. M., *Sov. Phys.-Semicond.*, 4, 124, 1970.

496. Kolomiets, B. T. and Raspopova, E. M., *Sov. Phys. Semicond.*, 4, 1041, 1971.

497. Kolomiets, B. T. and Lyubin, V. M., *Sov. Phys.-Solid State*, 2, 46, 1960.

498. Kolomiets, B. T. and Lyubin, V. M., *Sov. Phys.-Solid State*, 4, 291, 1962.

499. Mazets, T. F., Ph.D. Thesis, Leningrad Univ., 1969, unpublished.

500. Andriesh, A. M., Kolomiets, B. T., and Nazarova, T. F., *Sov. Phys.-Solid State*, 4, 1674, 1962.

501. Borisova, Z. U., *Izv. Akad. Nauk S.S.S.R., Ser Fiz*, 28, 1293, 1964.

502. Kruglov, V. I. and Bobrov, A. I., *Vestn. Leningrad Univ.*, 10, 125, 1966.

503. Kolomiets, B. T. and Rukhlyadev, Yu.V., *Sov. Phys.-Solid State*, 8, 2201, 1967.

504. Rose, A., *Concepts in Photoconductivity and Allied Problems*, Wiley-Interscience, New York, 1963.

505. Kolomiets, B. T. and Mazets, T. F., *J. Non-Crystalline Solids*, 3, 46, 1970.

506. Bube, R. H. and Thomson, S. M., *J. Chem. Phys.*, 23, 18, 1955.

507. Randal, J. T. and Wilkins, M. H. F., *Proc. Roy. Soc.*, (London) A184, 366, 1945.

508. Grossweiner, L. J., *J. Appl. Phys.*, 24, 1306, 1953.

509. Garlick, G. J. F. and Gibson, A. F., *Proc. Roy. Soc.*, (London) A60, 574, 1948.

510. Kolomiets, G. T. Mamontova, T. N., and Babaev, A. A., *J. Non-Crystalline Solids*, 4, 289, 1970.

511. Kolomiets, B. T., Mamontova, T. N., and Negreskul, V. V., *Phys. Status Solidi*, 27, K15, 1968.

512. Fischer, R., Heim, U., Stern, F., and Weiser, K., *Phys. Rev. Lett.*, 26, 1182, 1971.

513. Weiser, K., Fischer, R., and Brodsky, M. H., *Proc. Tenth Int. Conf. Phys. Semicond., Cambridge, Mass., 1970*, U.S.A.E.C. Div. Tech. Inform., Oak Ridge, Tenn., 1970, 667.

514. Kolomiets, B. T., *Proc. Ninth Int. Conf. Phys. Semicond., Moscow, 1968*, Nauka, Leningrad, 1968. 1259.

515. Kruglov, V. I. and Zimkina, T. M., *Sov. Phys.-Solid State*, 10, 170, 1968.

516. Glaze, F. W., Blackburn, D. H., Osmalov, J. S., Hubbard, D., and Black, M. H., *J. Res. N. B. S.*, 59, 83, 1957.

517. Minomura, S., Matsuda, J., and Oura, M., *Rev. Phys. Chem. Jap.*, 29, 22, 1959.

518. Kolomiets, B. T. and Pavlov, B. V., *Sov. Phys.-Solid State*, 2, 592, 1960.

519. Kosek, F. and Tauc, J., *Czech. J. Phys.*, B20, 94, 1970.

520. Andreichin, R., *J. Non-Crystalline Solids*, 4, 73, 1970.

521. Young, P. A., *Appl. Optics*, 10, 222, 1971.

522. Andrievskii, A. I., Nabitovich, I. D., and Voloshchuk, Ya.V., *Sov. Phys.-Cryst.* 6, 534, 1962.

523. Rockstad, H. K., *Solid State Communications,* 7, 1507, 1969.

524. Weiser, K. and Brodsky, M. H., *Phys. Rev.*, B1, 791, 1970.

525. Coutts, M. D. and Levin, E. R., *J. Appl. Phys.*, 38, 4039, 1967.

526. Lyubin, V. M. and Maidzinskii, V. S., *Sov. Phys.-Solid State*, 6, 3001, 1965.

527. Lyubin, V. M. and Maidzinskii, V. S., *Sov. Phys.-Semicond.*, 3, 1430, 1971.

528. Tatarinova, L. I., *Sov. Phys.-Cryst.*, 4, 637, 1959.

529. Ruby, S. L., Gilbert, L. R., and Wood, C., *Bull. Amer. Phys. Soc.*, 16, 303, 1971.

530. Schubert, K. and Fricke, H., *Z. Naturforsch.*, 6a, 781, 1951.

531. Schubert, K. and Fricke, H., *Z. Metallk.*, 44, 457, 1953.

532. Goldak, J., Barrett, G. S., Innes, D., and Youdelis, W., *J. Chem. Phys.*, 44, 3323, 1966.

533. McHugh, J. P. and Tiller, W. A., *Trans. AIME*, 218, 188, 1960.

534. Wooley, J. C. and Nikolic, P., *J. Electrochem. Soc.*, 112, 82, 1965.

535. Cohen, M. H., Falicov, L. M., and Golin, S., *IBM J. Res. Develop.*, 8, 215, 1964.

536. Hume-Rothery, W., *Electrons, Atoms, Metals, and Alloys*, Philosophical Library, New York, 1955.

537. Bierly, J. N., Muldawer, L., and Beckman, O., *Acta. Met.*, 11, 447, 1963.

538. Phillips, J. C., *Rev. Mod. Phys.*, 42, 317, 1970.

539. Adler, D., Cohen, M. H., Fagen, E. A., and Thompson, J. C., *J. Non-Crystalline Solids,* 3, 402, 1970.

540. Abrukisov, N. K., Vasserman, A. M., and Poretskaia, L. V., *Proc. Acad. Sci. U.S.S.R., Chem. Soc.*, 123, 817, 1958.

541. Hilton, A. R., Jones, C. E., and Brau, M., *Phys. Chem. Glasses*, 7, 105, 1966.

542. Howard, W. E. and Tsu, R., *Bull. Amer. Phys. Soc.*, 14, 428, 1969.

543. Chopra, K. L. and Bahl, S. K., *Bull. Amer. Phys. Soc.*, 14, 98, 1969.

544. Chopra, K. L. and Bahl, S. K., *J. Appl. Phys.*, 40, 4171, 1969.

545. Duwez, P., *Progress in Solid State Chemistry*, Reiss, H., Ed., Pergamon Press Oxford, 1967, 3, 377.

546. Bahl, S. K. and Chopra, K. L., *J. Vac. Sci. Technol.*, 6, 561, 1969.

547. Bienenstock, A., Betts, F., and Ovshinsky, S. R., *J. Non-Crystalline Solids*, 2, 347, 1970.

548. Betts, F., Bienenstock, A., and Ovshinsky, S. R., *J. Non-Crystalline Solids*, 4, 554, 1970.

549. Dove, D. B., Heritage, M. B., Chopra, K. L., and Bahl, S. K., *Appl. Phys. Lett.*, 16, 138, 1970.

550. Luo, H. L., Ph.D. Thesis, California Institute of Technology, 1964, unpublished.

551. Hilton, A. R., Jones, C. E., Dobrott, R. D., Klein, H. M., Bryant, A. M., and George, T. D., *Phys. Chem. Glasses*, 4, 116, 1966.

552. Parthe, E., *Crystal Chemistry of Tetrahedral Structures*, Gordon and Breach, New York, 1964.

553. Betts, F.,Bienenstock, A. I., and Bates, C. W., *Bull. Amer. Phys. Soc.*, 15, 1616, 1970.

554. Bienenstock, A. I., Betts, F., Keating, D. T., and deNeufville, J., *Bull. Amer. Phys. Soc.*, 15, 1616, 1970.

555. Ure, R. W., Bowers, R., and Miller, R. C., *Properties of Elemental and Compound Semiconductors*, Gatos, H. C., Ed., Interscience, New York, 1960, 254.

556. Mazelsky, R. and Lubell, M. S., *Adv. Chem.*, 39, 210, 1963.

557. Kolomoets, N. V., Lev, E. Ya., and Sysoeva, L. M., *Sov. Physics-Solid State*, 5, 2101, 1964.

558. Kolomoets, N. V., Lev, E. Ya., and Sysoeva, L. M., *Sov. Phys.-Solid State*, 6, 551, 1964.

559. Tsu, R., Howard, W. E., and Esaki, L., *Solid State Commun.*, 5, 167, 1967.

560. Logachev, Yu. A. and Moizhes, B. Ya., *Inorg. Mater.*, 6, 1577, 1971.

561. Brebrick, B. F., *J. Phys. Chem. Solids*, 27, 1495, 1966.

562. Bahl, S. K. and Chopra, K. L., *J. Appl. Phys.*, 41, 2196, 1970.

563. Esaki, L., *J. Phys. Soc. Jap. Suppl.*, 21, 589, 1966.

564. Stiles, P. J., Esaki, L., and Howard, W. E., *Proc. Tenth Int. Conf. Low Temperature Phys., Moscow, 1966*, VINITI, Moscow, 1967, 257.

565. Tsu, R., Howard, W. E., and Esaki, L., *Phys. Rev.*, 172, 779, 1968.

566. Burstein, E., *Phys. Rev.*, 93, 632, 1954.

567. Erasova, N. A., Kaidanov, V. I., Chernik, I. A., Sysoeva, L. M., Lev, E. Ya., and Kolomoets, N. V., *Sov. Phys.-Semicond.*, 3, 1075, 1970.

568. Tung, Y. W. and Cohen, M. L., *Phys. Rev.*, 180, 823, 1969.

569. Hein, R. A., Gibson, J. W., Mazelsky, R., Miller, R. C., and Hulm, J. K., *Phys. Rev. Lett., 12, 320, 1964.*

570. Howard, W. E. and Tsu, R., *Phys. Rev.*, B1, 4709, 1970.

571. See, for example, Adler, D., *Solid State Phys.*, 21, 59, 1968.

572. Bahl. S. K. and Chopra, K. L., *J. Appl. Phys.*, 40, 4940, 1969.

573. Howard, W. E. and Tsu, R., *Proc. Tenth Int. Conf. Phys. Semicond., Cambridge, Mass., 1970*, U.S.A.E.C. Div. Tech. Inform., Oak Ridge, Tenn., 1970, 789.

574. Schultz-Sellack, C., *Ann. Phys. Chem.*, 139, 182, 1870.

575. Frerichs, R., *J. Opt. Soc. Amer.*, 43, 1153, 1953.

576. Fraser, W. A., *J. Opt. Soc. Amer.*, 43, 823, 1953.

577. Glaze, F. W., Blackburn, D. H., Osmalov, J. S., Hubbard, D., and Black, M. H., *J. Res. N. B. S.*, 59, 83, 1957.

578. Goryunova, N. A. and Kolomiets, B. T., *Zh. Tech. Fiz.*, 25, 984, 1955.

579. Goryunova, N. A. and Kolomiets, B. T., *Izv. Akad. Nauk. S.S.S.R.*, Ser. Fiz., 20, 1496, 1956.

580. Kolomiets, B. T. and Nazarova, T. F., *Sov. Phys.-Solid State*, 2, 369, 1960.

581. Goryunova, N. A., Shylo, V. P., and Kolomiets, B. T., *Sov. Phys.-Tech. Phys.*, 3, 912, 1958.

582. Flaschen, S. S., Pearson, A. D., and Northover, W. R., *J. Amer. Ceram. Soc.*, 42, 450, 1959.

583. Nazarova, T. F. and Kolomiets, B. T., *Fiz. Tverd. Tela.*, 2, 22, 1959.

584. Flaschen, S. S., Pearson, A. D., and Northover, W. R., *J. Amer. Ceram. Soc.*, 43, 274, 1960.

585. Goryunova, N. A. and Kolomiets, B. T., *Vitreous State*, Izd. Akad. Nauk S.S.S.R., 1960.

586. Kolomiets, B. T., *Vitreous State*, Izd. Akad. Nauk S.S.S.R., 1960.

587. Goryunova, N. A., Kolomiets, B. T., and Shilo, V. P., *Sov. Phys.-Solid State*, 2, 258, 1960.

588. Kolomiets, B. T. and Nazarova, T. F., *Sov. Phys.-Solid State*, 2, 159, 1960.

589. Muller, R. L., *Vitreous State*, Izd. Akad. Nauk S.S.S.R., 1960, 61.

590. Ayo, L. G., and Korkorina, V. F., *Optiko-Mekhanicheskaya Promyshlenmost*, 4, 1961.

591. Baydokov, L. A., Borisova, Z. U., and Muller, R. L., *Zh. Prikl. Khimii*, 34, 2446, 1961.

592. Ayo, L. G. and Kokorina, V. F., *Optiko-Mekhanicheskaya Promysehlenmost*, 6, 1961.

593. Muller, R. L., Orlova, G. M., Timofeejeva, V. N., and Ternovaya, G. I., *Vestn. Leningrad Univ.*, 22, 146, 1962.

594. Muller, R. L. and Markova, T. P., *Vestn. Leningrad Univ.*, 22, 75, 1962.

595. Baydakov, L. A., *Vestn. Leningrad Univ.*, 22, 159, 1962.

596. Borisova, Z. U. and Bobrov, A. T., *Vestn. Leningrad Univ.*, 22, 159, 1962.

597. Markova, T. P., *Vestn. Leningrad Univ.*, 22, 96, 1962.

598. Egorova, E. A. and Kokorina, V. F., *Optiko-Mekhanicheskaya Promyshlenmost*, 1, 1963.

599. Ayo, L. G. and Kokorina, V. F., *Optiko-Mekhanicheskaya Promyshlenmost*, 2, 1963.

600. Hilton, A. R. and Brau, M., *Infrared Phys.*, 3, 69, 1963.

601. Hilton, A. R., Jones, C. E., and Brau, M., *Infrared Phys.*, 4, 213, 1964.

602. Pearson, A. D., *Modern Aspects of the Vitreous State*, Vol. 3, Mackenzie, J. D., Ed., Butterworths, London, 1964, 29.

603. Flaschen, S. S., Pearson, A. D., and Northover, W. R., *J. Appl. Phys.*, 31, 219, 1960.

604. Pearson, A. D., Northover, W. R., Dewald, J. F., and Peck, W. F., *Advances in Glass Technology*, Part I, Plenum, New York, 1962, 357.

605. Lin, F. C. and Ho, S. M., *J. Amer. Ceram. Soc.*, 46, 24, 1963.

606. Meyerowitz, R., *Amer. Mineral*, 40, 398, 1955.

607. Fischer, A. G. and Mason, A. S., *J. Opt. Soc. Amer.*, 52, 721, 1962.

608. Petz, J. I., Kruh, R. F., and Amstatz, G. C., *J. Chem. Phys.*, 34, 526, 1961.

609. Hopkins, T. E., Pasternack, R. A., Gould, E. S., and Herndon, J. R., *J. Phys. Chem.*, 66, 733, 1962.

610. Litvinova, S. M., Misselyuk, E. G., Tyurik, Y. A., Shylo, V. P., and Kolomiets, B. T., *Radioteckhnika Electrotekhnika*, 7, 1054, 1962.

611. Dembovkii, S. A. and Vaipolin, A. A., *Sov. Phys.-Solid State*, 6, 1388, 1964.

612. Domorad, I. A. and Khiznichenko, L. P., *Izv. Akad, Nauk S.S.S.R., Ser. Fiz.*, 4, 99, 1963.

613. Domorad, I. A., Kaipnazarov, K., and Khiznichenko, L. P., *Izv. Akad. Nauk S.S.S.R.,Ser. Fiz.*, 5, 87, 1963.

614. Hilton, A. R., *Appl. Opt.*, 5, 1977, 1966.

615. Tsuei, C. C. and Kankeleit, E., *Phys. Rev.*, 162, 312, 1967.

616. Senturia, S. D., Hewes, C. R., and Adler, D., *J. Appl. Phys.*, 41, 430, 1970.

617. Muller, R. L., *J. Appl. Chem. U.S.S.R.*, 35, 519, 1962.

618. Fritzsche, H., *Electronic Properties of Materials*, Bube, R. H., Ed., McGraw-Hill, New York, in press.

619. Minami, T., Hattori, M., Nakamachi, F., and Tanaka, M., *J. Non-Crystalline Solids*, 3, 327, 1970.

620. Haisty, R. W. and Krebs, H., *J. Non-Crystalline Solids*, 1, 399, 1969.

621. Krebs, H., *J. Non-Crystalline Solids*, 1, 455, 1969.

622. Male, J. C., *Electronics Lett.*, 6, 91, 1970.

623. Adler, D., *Solid State Phys.*, 21, 97, 1968.

624. Fagen, E. A. and Fritzsche, H., *J. Non-Crystalline Solids*, 2, 170, 1970.

625. Croitoru, N., Vescan, L., Popescu, C., and Lazarescu, M., *J. Non-Crystalline Solids*, 4, 493, 1970.

626. Croitoru, N., Vescan, L., Botila, T., Vancu, A., Lazarescu, M., and Ioanid, G., *Proc. Congres Int. sur les Couches Minces, Cannes, 1970*, S.F.I.T.V., Paris, 1970, 455.

627. Uphoff, H. L. and Healy, J. H., *J. Appl. Phys.*, 33, 2770, 1962.

628. Quinn, R. K. and Johnson, R. T., *Bull. Amer. Phys. Soc.*, 16, 304, 1971.

629. Deis, D. W. and Dancy, E. A., *Bull. Amer. Phys. Soc.*, 16, 303, 1971.

630. Hilborn, R. B. and Prasad, K., *J. Vac. Sci. Technol.*, 6, 632, 1969.

631. Fagen, E. A., Holmberg, S. H., Seguin, R. W., Thompson, J. C., and Fritzsche, H., *Proc. Tenth Int. Conf. Phys. Semicond., Cambridge, Mass., 1970*, U.S.A.E.C. Div. Tech. Inform., Oak Ridge, Tenn., 1970, 672.

632. Tsuei, C. C., *Phys. Rev.*, 170, 775, 1968.

633. de Neufville, J., First Semi-Annual Technical Report, Contract DAHC15-70-C-0187, 1970, unpublished.

634. Rockstad, H. K., Flasck, R. A., and Iwasa, S., *J. Non-Crystalline Solids*, in press.

635. Adler, D., *Solid State Phys.*, 21, 70, 1968.

636. Nagels, P., Callaerts, R., Denayer, M., and DeConinck, R., *J. Non-Crystalline Solids*, 4, 295, 1970.

637. Peck, W. F. and Dewald, J. F., *J. Electrochem. Soc.*, 111, 561, 1964.

638. Male, J. C., *Brit. J. Appl. Phys.*, 18, 1543, 1967.

639. Ivkin, E. B., Kolomiets, B. T., and Lebedev, E. A., *Bull. Acad. Sci. U.S.S.R.*, Phys. Ser. 28, 1190, 1965.

640. Kornfel'd, M. I. and Sochava, L. S., *Sov. Phys.-Solid State*, 1, 1256, 1960.

641. Friedman, L. and Holstein, T., *Ann. Phys.*, (N.Y.) 21, 494, 1963.

642. Holstein, T. and Friedman, L., *Phys. Rev.*, 165, 1019, 1968.

643. Pearson, A. D., *J. Electrochem. Soc.*, 111, 753, 1964.

644. Allgaier, R. S., *Proc. Ninth Int. Conf. Phys. Semicond., Moscow, 1968*, Nauka, Leningrad, 1968, 1271.

645. Allgaier, R. S., *J. Vac. Sci. Technol.*, 8, 113, 1971.

646. Boer, K. W., *J. Non-Crystalline Solids*, 2, 444, 1970.

647. Boer, K. W., *Phys. Status Solidi*, 34, 721, 1969.

648. Van Daal, H. J. and Bosman, A. J., *Phys. Rev.*, 158, 736, 1967.

649. Tieche, Y. and Zareba, A., *Phys. Kondens. Materie*, 1, 402, 1963.

650. Allgaier, R. S., *Phys. Rev.*, 185, 227, 1969.

651. Maranzana, F. E., *Phys. Rev.*, 160, 421, 1967.

652. Banyai, L. and Aldea, A., *Phys. Rev.*, 143, 652, 1966.

653. Jones, R., *J. Phys.*, C, 3, 202, 1970.

654. Rockstad, H. K., *J. Non-Crystalline Solids*, 2, 192, 1970.

655. Bishop, S. G., Taylor, P. C., Mitchell, D. L., and Slack, L. H., *J. Non-Crystalline Solids*, 5, 351, 1971.

656. Rockstad, H. K., *Bull. Amer. Phys. Soc.*, 16, 304, 1971.

657. Fritzsche, H., Fagen, E. A., and Ovshinsky, S. R., unpublished; quoted in ref. 658.

658. Ovshinsky, S. R., Evans, E. J., Nelson, D. L., and Fritzsche, H., *IEEE Trans. Nucl. Sci.*, NS-15, 311, 1968.

659. Barbe, D. F., *J. Vac. Sci. Technol.*, 8, 102, 1971.

660. Heilmeier, G. G. and Zanoni, L. A., *J. Phys. Chem. Solids*, 25, 603, 1964.

661. Tick, P. A. and Hindley, N. K., *Bull. Amer. Phys. Soc.*, 16, 304, 1971.

662. Black, J., Conwell, E., Seigle, Z., and Spencer, Z., *Phys. Chem. Solids*, 2, 240, 1957.

663. Kolomiets, B. T. and Pavlov, B. V., *Sov. Phys.-Solid State*, 2, 592, 1960.

664. Pavlov, B. V. and Kolomiets, B. T., *Vitreous State*, Izd. Akad. Nauk S.S.S.R., 1960.

665. Vashko, A., Prokopova, G., Pavlov, E., Shylo, V. P., and Kolomiets, B. T., *Optika i Spektroskopiya*, 12, 275, 1962.

666. Fagen, E. A. and Fritzsche, H., *J. Non-Crystalline Solids*, 2, 180, 1970.

667. Fagen, E. A., unpublished data.

668. Boer, K. W., Hansch, H. J., and Kummel, V., *Z. Phys.*, 155, 170, 1959.

669. Williams, R., *Phys. Rev.*, 117, 1487, 1960.

670. Andriesh, A. M. and Kolomiets, B. T., *Sov. Phys.-Solid State*, 5, 1063, 1963.

671. Botila, T. and Vancu, A., *Mat. Res. Bull.*, 5, 925, 1970.

672. Fagen, E. A. and Fritzsche, H., *J. Non-Crystalline Solids*, 4, 480, 1970.

673. Fagen, E. A., *Bull. Amer. Phys. Soc.*, 15, 331, 1970.

674. Arnoldussen, T., Bube, R. H., and Fagen, E. A., *J. Non-Crystalline Solids*, in press.

675. Rose, A., *Concepts in Photoconductivity and Allied Problems*, Interscience, New York, 1963.

676. Shaw, R. F., Ph.D. Thesis, Cambridge Univ., 1969, unpublished.

677. Keyes, R. J., *J. Appl. Phys.*, 38, 2619, 1967.

678. Morrison, S. R., *Phys. Rev.*, 104, 619, 1956.

679. Haynes, J. R. and Hornbeck, J. A., *Phys. Rev.*, 100, 606, 1955.

680. Dobrego, V. P. and Ryvkin, S. M., *Sov. Phys.-Solid State*, 6, 928, 1964.

681. Dobrego, V. P., Ryvkin, S. M., and Shkol'nik, A. L., *Sov. Phys.-Solid State*, 7, 671, 1965.

682. Fuhs, W. and Stuke, J., *Phys. Status Solidi*, 27, 171, 1968.

683. Drabkin, I. A., Emel'yanova, L. T., Iskenderov, R. N., and Ksendzov, Ya.M., *Sov. Phys.-Solid State*, 10, 2428, 1969.

684. Gregory, B. L., *Appl. Phys. Lett.*, 16, 67, 1970.

685. Boer, K. W., Esbitt, A. S., and Kaufman, W. M., *J. Appl. Phys.*, 37, 2664, 1966.

686. Berzelius, J. J., *Ann. Phys. Chem.*, 32, 577, 1834.

687. Roscoe, H. E., *Phil. Trans. Roy. Soc.*, (London) 158, 1, 1868.

688. Lehner, V. and Wolesensky, E., *J. Amer. Chem. Soc.*, 35, 718, 1913.

689. Tamman, G. and Jenckel, E., *Z. Anorg. Allg. Chem.*, 184, 416, 1929.

690. Sheperd, E. S., Rankin, G. A., and Wright, F. E., *Amer. J. Sci.*, 28, 293, 1909.

691. Rankin, G. A. and Merwin, H. E., *J. Amer. Chem. Soc.*, 38, 568, 1916.

692. Akiyama, K. I., and Sawayama, G., Memo. Faculty Sci. Engin., Waseda Univ., 11, 173, 1934.

693. Bussem, W. and Eitel, W., *S. Krist.*, 95, 175, 1936.

694. Eitel, W. and Skaliks, W., *Z. Anorg. Allg. Chem.*, 183, 263, 1929.

695. Baynton, P. L., Rawson, H., and Stanworth, J. E., *Proc. Fourth Int. Congr. Glass, Paris, 1956*, Imprimerie Chaix, Paris, 1956, 52.

696. Janakirama-Rao, Bh.V., *J. Amer. Ceram. Soc.*, 47, 455, 1964.

697. Forland, J. and Weyl, W. A., *J. Amer. Ceram. Soc.*, 33, 186, 1950.

698. Rostkowsky, A. P., *Zh. Russk. Fiz. Khim. Obshch.*, 62, 2055, 1930.

699. Bergman, A. G., *Doklady Akad. Nauk S.S.S.R.*, 38, 304, 1943.

700. Stevels, J. M., *Philips Tech. Rev.*, 13, 293, 1952.

701. Baynton, P. L., Rawson, H., and Stanworth, J. E., *Nature*, 179, 434, 1957.

702. Franck, H. H., *Tag. Ber. Chem. Ges. D.D.R.*, 119, 1955.

703. Rothermel, J. J., Sun, K. H., and Silverman, A., *J. Amer. Ceram. Soc.*, 32, 153, 1949.

704. Sun, K. H., *Glass Ind.*, 27, 552, 1946.

705. Schroder, J., *Angew. Chem. Int. Ed.*, 3, 376, 1964.

706. Maier, C. G., U.S. Bureau of Mines Tech. Paper No. 360, 1925, unpublished.

707. Mackenzie, J. D. and Murphy, W. K., *J. Chem. Phys.*, 33, 366, 1960.

708. Secrist, D. R. and Mackenzie, J. D., *Advances in Chemistry*, No. 80, 1969.

709. Mickelsen, R. A. and Kingery, W. D., *J. Appl. Phys.*, 37, 3541, 1966.

710. Frey, W. J., Ph.D. Thesis, R. P. I., 1969, unpublished.

711. Mackenzie, J. D., *Proc. Congress Int. sur les Couches Minces, Cannes, 1970*, S.F.I.T.V., Paris, 1970, 427.

712. Barbe, D. F. and Herman, D. S., *J. Appl. Phys.*, 41, 3116, 1970.

713. Shepard, K. W., *J. Appl. Phys.*, 36, 796, 1965.

714. Morey, G. W., *Properties of Glass*, Reinhold, New York, 1954.

715. Denton, E. P., Rawson, H., and Stanworth, J. E., *J. Electrochem. Soc.*, 104, 237, 1957.

717. Munakata, M., *Solid State Electronics*, 1, 159, 1960.

718. Nester, H. H. and Kingery, W. D., *Proc. Seventh Int. Congr. Glass, Brussels, 1965*, Gordon and Breach, New York, 1965, 106.

719. Mackenzie, J. D., *Bull. Amer. Ceram. Soc.*, 42, 223, 1963.

720. Ioffe, V. A., Patrina, J. B., and Poberovshaya, I. S., *Sov. Phys.-Solid State*, 2, 609, 1960.

721. Munakata, M. and Iwamoto, M., *J. Ceram. Ass. Jap.*, 68, 125, 1960.

722. Grechanik, L. A., Petrovykh, N. V., and Karpechenko, V. G., *Sov. Phys.-Solid State*, 2, 1908, 1961.

723. Hamblen, D. P., Weidel, R. A., and Blair, G. E., *J. Amer. Ceram. Soc.*, 46, 499, 1963.

724. McMillan, P. W., *Advances in Glass Technology*, Vol. 1, Plenum Press, New York, 1963, 333.

725. Mazurin, O. V., Pavlova, G. A., Lev, E.Ya., and Leko, E. K., *Z. Tech. Phys. U.S.S.R.*, 27, 2702, 1957.

726. Evstrop'yev, K. K. and Tsekhomskii, V. A., *Sov. Phys.-Solid State*, 4, 2482, 1962.

727. Trap, H. J. L. and Stevels, J. M., *Verres Refractaires*, 16, 337, 1962.

728. Grechanik, L. A., Fainberg, E. A., and Zertsalova, I. N., *Sov. Phys.-Solid State,* 4, 331, 1962.

729. Warren, B. E. and Biscoe, J., *J. Amer. Ceram. Soc.*, 21, 259, 1938.

730. Warren, B. E. and Biscoe, J., *J. Amer. Ceram. Soc.*, 21, 49, 1938.

731. Biscoe, J. and Warren, B. E., *J. Amer. Ceram. Soc.*, 21, 287, 1938.

732. Porai-Koshits, E. A., *Proc. Conf. Structure Glass, Leningrad, 1953*, Consultants Bureau, New York, 1958, 25.

733. Warren, B. E., Krutter, H., and Morningstar, O., *J. Amer. Ceram. Soc.*, 19, 202, 1936.

734. Herre, F. and Richter, H., Z. Naturforsch., 12A, 545, 1957.

735. Biscoe, J., Pincus, A. G., Smith, C. S., and Warren, B. E., *J. Amer. Ceram. Soc.*, 24, 116, 1941.

736. Brady, G. W., *J. Chem. Phys.*, 28, 48, 1958.

737. Brady, G. W., *J. Chem. Phys.*, 24, 477, 1956.

738. Brady, G. W., *J. Chem. Phys.*, 27, 300, 1957.

739. Plieth, K., Reuber, E., and Stranski, I. N., *Z. Anorg. Allg. Chem.*, 280, 205, 1955.

740. Bachmann, H. G., Ahmed, F. R., and Barnes, W. H., *Z. Krist.*, 115, 110, 1961.

741. Hill, W. L., Faust, G. T., and Hendricks, S. B., *J. Amer. Chem. Soc.*, 65, 794, 1943.

742. deDecker, J. C. J. and MacGillavry, C. H., *Recl. Trav. Chim. Pays-Bas Belg.*, 60, 153, 1941.

743. deDecker, H. C. J., *Recl. Trav. Chim. Pays-Bas Belg.*, 60, 413, 1941.

744. MacGillavry, C. H., deDecker, H. C. J., and Nijland, L. M., *Nature*, 164, 448, 1949.

745. Sadagopan, V. and Gatos, H. C., *Mater. Sci. Eng.*, 2, 273, 1967.

746. Janakirama-Rao, Bh.V., *J. Amer. Ceram. Soc.*, 48, 311, 1965.

747. Janakirama-Rao, Bh.V., *J. Amer. Ceram. Soc.*, 49, 605, 1966.

748. Anderson, G. W. and Compton, W. D., *J. Chem. Phys.*, 52, 6166, 1970.

749. Kenny, N., Kannewurf, C. R., and Whitmore, *J. Phys. Chem. Solids*, 27, 1237, 1966.

750. Bodo, Z. and Hevesi, I., *Phys. Status Solidi*, 20, K45, 1967.

751. Hevesi, I., *Acta Phys. Acad. Sci. Hung.*, 23, 75, 1967.

752. Hevesi, I., *Acta Phys. Acad. Sci. Hung.*, 23, 415, 1967.

753. Hevesi, I., *Acta Phys. Chem. Szeged.*, 13, 39, 1967.

754. Boros, J., *Z. Phys.*, 126, 721, 1949.

755. Conlon, D. C. and Doyle, W. P., *J. Chem. Phys.*, 35, 752, 1961.

756. Miller, F. A. and Baer, W. K., *Spectrochim. Acta*, 17, 112, 1961.

757. Sathyanarayana, D. N. and Patel, C. C., *J. Inorg. Nucl. Chem.*, 30, 207, 1968.

758. Manhanti, P. C., *Proc. Phys. Soc.*, London, 47, 433, 1935.

759. Vratny, F., *J. Inorg. Nucl. Chem.*, 21, 77, 1961.

760. Curry, J., Herzberg, L., and Herzberg, G., *Z. Phys.*, 86, 348, 1933.

761. Daasch, L. W., and Smith, D. C., *Anal. Chem.*, 23, 853, 1951.

762. Gosh, P. N. and Ball, G. N., *Z. Phys.*, 71, 362, 1961.

763. Butcher, F. K., Deuters, B. E., Gerrard, W., Mooney, E. F., Rothenbury, R. A., and Willis, H. A., *Spectrochim. Acta*, 20, 759, 1964.

764. Chapman, A. C. and Thirlwell, L. E., *Spectrochim. Acta*, 20, 937, 1964.

765. Corbridge, D. E. C. and Lowe, E. J., *J. Chem. Soc.*, 1954, 493, 1954.

766. Corbridge, D. E. C. and Lowe, E. J., *Anal. Chem.*, 27, 1383, 1955.

767. Bues, W. and Gehrke, H. W., *Z. Anorg. Allg. Chem.*, 288, 291, 1956.

768. Williams, D. J., Bradbury, B. T., and Maddocks, W. R., *J. Soc. Glass Tech.*, 43, 337, 1959.

769. Landsberger, F. R. and Bray, P. J., *J. Chem. Phys.*, 53, 2757, 1970.

770. Gornostansky, S. D. and Stager, C. V., *J. Chem. Phys.*, 46, 4959, 1967.

771. Baugher, J. F., Taylor, P. C., Oja, T., and Bray, P. J., *J. Chem. Phys.*, 50, 4914, 1967.

772. Nador, B., *Glass Ceram.*, 17, 517, 1960.

773. Brown, R. M., Ph.D. Thesis, U. of Illinois, 1966.

774. Mackenzie, J. D., Tech. Rep. 5, Contract Nonr-591 (21).

775. Kennedy, T. N. and Mackenzie, J. D., *Phys. Chem. Glasses*, 8, 169, 1967.

776. Schmid, A. P., *J. Appl. Phys.*, 39, 3140, 1968.

777. Linsley, G. S., Owen, A. E., and Hayatee, F. M., *J. Non-Crystalline Solids*, 4, 208, 1970.

778. Munakata, M., Kawamura, S., Asahara, J., and Iwamoto, M., *Yogyo Kyokai Shi*, 67, 344, 1959.

779. Mackenzie, J. D., *Modern Aspects of the Vitreous State*, Vol. 3, Mackenzie, J. D., Ed., Butterworths, Washington, D. C., 1964, 126.

780. Allersma, T. and Mackenzie, J. D., *J. Chem. Phys.*, 47, 1406, 1967.

781. Hakim, R., M. S. Thesis, R. P. I., 1965, unpublished.

782. Schmid, A. P., *J. Non-Crystalline Solids*, 4, 232, 1970.

783. Klinger, M. I., *Phys. Status Solidi*, 27, 479, 1968.

784. Friedman, L., *Phys. Rev.*, 135, A233, 1964.

785. Mott, N. F., *J. Non-Crystalline Solids*, 1, 1, 1969.

786. Springthrope, A. J., Austin, I. G., and Austin, B. A., *Solid State Commun.*, 3, 143, 1965.

787. Appel, J. *Solid State Phys.*, 21, 323, 1968.

788. Ohashi, S. and Matsumura, T., *Bull. Chem. Soc. Jap.*, 35, 501, 1962.

789. Heikes, R. R. and Ure, R. W., *Thermoelectricity*, Interscience, New York, 1961, 81.

790. Sewell, G. L., *Phys. Rev.*, 129, 597, 1963.

791. Schotte, K. D., *Z. Phys.*, 196, 393, 1966.

792. Efros, A. L., *Sov. Phys.-Solid State*, 9, 901, 1967.

793. Klinger, M. I., *Rep. Progr. Phys.*, 30, 225, 1968.

794. Ohashi, S., *Topics in Phosphorous Chemistry*, Vol. 1, Interscience, New York, 1964, 217.

795. Linsley, G. S., Ph.D. Thesis, Sheffield Univ., 1969.

796. Wyckoff, R. W. G., *Crystal Structures*, Vol. 2, John Wiley & Sons, New York, 1963, 187.

797. Adler, D., *Rev. Mod. Phys.*, 40, 714, 1968.

798. Allersma, T., Hakim, R. H., Kennedy, T. N., and Mackenzie, J. D., *J. Chem. Phys.*, 46, 154, 1967.

799. Volzhenskii, D. S. and Pashkovskii, M. V., *Sov. Phys.-Solid State*, 11, 950, 1969.

800. Nagels, P. and Denayer, M., *Proc. Tenth Int. Conf. Phys. Semicond., Cambridge, Mass., 1970*, U.S.A.E.C.Div. Tech. Inform., Oak Ridge, Tenn., 1970, 321.

801. Ioffe, V. A. and Patrina, I. B., *Proc. Ninth Int. Conf. Phys. Semicond., Moscow, 1968*, Nauka, Leningrad, 1968, 1151.

802. Patrina, I. B. and Ioffe, V. A., *Sov. Phys.-Solid State*, 6, 2581, 1964.

803. Ioffe, V. A. and Patrina, I. B., *Sov. Phys.-Solid State*, 6, 2425, 1964.

804. Dmitrieva, L. V., Ioffe, V. A., and Patrina, I. B., *Sov. Phys.-Solid State*, 7, 2228, 1966.

805. Adler, D., *Critical Phenomena*, Mills, R. E., Ascher, E., and Jaffee, R. I., Eds., McGraw-Hill, New York, 1971, 567.

806. Rice, T. M., McWhan, D. B., and Brinkman, W. F., *Proc. Tenth Int. Conf. Phys. Semicond., Cambridge, Mass., 1970*, U.S.A.E.C. Div. Tech. Inform., Oak Ridge, Tenn. 1970, 293.

807. Kennedy, T. N. and Mackenzie, J. D., *J. Non-Crystalline Solids*, 1, 326, 1969.

808. Mackenzie, J. D., personal communication.

809. Bardeen, J., Blatt, F. J., and Hatt, L., *Photoconductivity Conference*, Breckenridge, R. G., Russell, B. R., and Hahn, E. E., Eds., John Wiley & Sons, New York, 1956, 146.

810. Griffiths, J. S., *The Theory of Transition-Metal Ions*, Cambridge University Press, Cambridge, England, 1961.

811. Kramers, H. A., *Proc. Amsterdam Acad.*, 33, 959, 1930.

812. Roth, W. L., *Acta Crystallogr.*, 13, 140, 1960.

813. Smart, J. S., *Phys. Rev.*, 82, 113, 1951.

814. Ellefson, B. S. and Taylor, N. W., *J. Chem. Phys.*, 2, 58, 1934.

815. Hamilton, W. C., *Phys. Rev.*, 110, 1050, 1958.

816. Verwey, E. J. W., Haayman, P. W., and Romeijn, F. C., *J. Chem. Phys.*, 15 181, 1947.

817. Pauling, L. and Hendricks, S. B., *J. Amer. Chem. Soc.*, 47, 781, 1925.

818. Svendsen, M. B., *Naturwissenschaften,* 45, 542, 1958.

819. vanOosterhout, C. W. and Rooijman, C. J. M., *Nature*, 181, 44, 1958.

820. Hansen, K. W., *J. Electrochem. Soc.*, 112, 994, 1965.

821. Kinser, D. L., *J. Electrochem. Soc.*, 117, 546, 1970.

822. Maxwell, J. C., *Electricity and Magnetism*, Vol. 1, Clarendon Press, London, 1892, 328.

823. Wagner, K. W., *Arch. Elektrotech*, 2, 371, 1914.

824. Zener, C., *Proc. Roy. Soc.*, (London) 145, 523, 1934.

825. Fowler, R. H. and Nordheim, L. W., *Proc. Roy. Soc.*, (London) A119, 173, 1928.

826. Schottky, W., *Z. Phys.*, 15, 872, 1914.

827. Jonscher, A. K., *Thin Solid Films*, 1, 213, 1967.

828. Argall, F. and Jonscher, A. K., *Thin Solid Films*, 2, 185, 1968.

829. Jonscher, A. K., *J. Electrochem. Soc.*, 116, 217C, 1969.

830. Jonscher, A. K., *J. Vac. Sci. Tech.*, 8, 135, 1971.

831. Simmons, J. G., *J. Phys.*, D. 4, 613, 1971.

832. Simmons, J. G., *Handbook of Thin Film Technology*, Maissel, L., and Glang, H., Eds., McGraw-Hill, New York, 1970, Chap. 14.

833. Dearnaley, G., Stoneham, A. M., and Morgan, D. V., *Rep. Progr. Phys.*, 33, 1129, 1970.

834. Fritzsche, H., *I.B.M. J. Res. Dev.*, 13, 515, 1969.

835. Shaw, M. and Gastman, I. J., *Appl. Phys. Lett.*, to be published.

836. Weintraub, E., *J. Indus. Engin. Chem.*, 5, 106, 1913.

837. Lyle, F. W., *Phys. Rev.*, 11, 253, 1918.

838. Ovshinsky, S. R., unpublished research 1958, cited by Cooney, J. D., *Control Engineering*, 6, 121, 1959 and by Young, L., *Anodic Oxide Films*, Academic Press, New York, 1961, 147.

839. For a history of the early work, see Ovshinsky, S. R., *J. Non-Crystalline Solids*, 2, 99, 1970.

840. Ovshinsky, S. R., U.S. Patent 3,052,830, 1962.

841. Geppert, D. V., *Proc. IEEE*, 51, 223, 1963.

842. Chopra, K. L., *Proc. IEEE*, 51, 941, 1963.

843. Beam, W. R. and Armstrong, A. L., *Proc. IEEE*, 52, 300, 1964.

844. Chopra, K. L., *J. Appl. Phys.*, 36, 184, 1965.

845. Hayashi, T. and Niimi, T., *Proc. IEEE*, 52, 986, 1964.

846. Hayashi, T. and Nimmi, T., *Jap. J. Appl. Phys.*, 3, 500, 1964.

847. Laverty, S. J., *Int. J. Electronics*, 30, 165, 1971.

848. Pettus, C., *Proc. IEEE*, 53, 98, 1965.

849. Sawamura, K. and Saito, M., *Jap. J. Appl. Phys.*, 5, 182, 1966.

850. Dewald, J. F., Pearson, A. D., Northover, W. R., and Peck, W. F., *J. Electrochem. Soc.*, 109, 243C, 1962.

851. Pearson, A. D., Northover, W. R., Dewald, J. F., and Peck, W. F., *Advances in Glass Technology*, Matson, F. R., and Rindone, G. E., Eds., Plenum Press, New York, 1962, 357.

852. Kolomiets, B. T. and Lebedev, E. A., *Radio Eng. Electron. Phys.*, (U.S.S.R.) 8, 1941, 1963.

853. Eaton, D. L., *J. Amer. Ceram. Soc.*, 47, 533, 1964.

854. Ovshinsky, S. R., Fourth Symposium on Vitreous Chalcogenide Semiconductors, Leningrad, 1967, unpublished.

855. Walsh, P. J., Vogel, R., and Evans, E. J., *Phys. Rev.*, 178, 1274, 1969.

856. Pearson, A. D. and Miller, C. E., *Appl. Phys. Lett.*, 14, 280, 1969.

857. Stocker, H. J., *Appl. Phys. Lett.*, 15, 55, 1969.

858. Sugi, M., Kikuchi, M., Iizima, S., and Tanaka, K., *Solid State Commun.*, 7, 1805, 1969.

859. Haberland, D. R., Karmann, R., and Thoma, P., *Z. Angew. Phys.*, 28, 143, 1969.

860. Walsh, P. J., Hall, J. E., Nicolaides, R., Defeo, S., Calella, P., Kuchmas, J., and Doremus, W., *J. Non-Crystalline Solids*, 2, 107, 1970.

861. Hall, J. E., *J. Non-Crystalline Solids*, 2, 125, 1970.

862. Csillag, A. and Jager, H., *J. Non-Crystalline Solids*, 2, 133, 1970.

863. Deis, D. W., Dancy, E. A., and Patterson, A., *J. Non-Crystalline Solids*, 2, 141, 1970.

864. Haden, C. R., Stone, J. L., Linder, J. S., *Proc. IEEE*, 58, 1852, 1970.

865. Armitage, D., Brodie, D. E., and Eastman, P. C., *Canad. J. Phys.*, 48, 2780, 1970.

866. Haberland, D. R. and Kehrer, H. P., *Solid-State Electronics*, 13, 451, 1970.

867. Haberland, D. R., *Solid-State Electronics*, 13, 207, 1970.

868. Haberland, D. R., *Sonderdruck Nachrichtentechnische Z.*, 9, 449, 1970.

869. Haberland, D. R., Karmann, R., and Repp, F., *Sonderdruck Z. Frequenz*, 7, 212, 1970.

870. Haberland, D. R., *Sonderdruck Z. Frequenz*, 6, 185, 1970.

871. Weirauch, D. F., *Appl. Phys. Lett.*, 16, 72, 1970.

872. Stone, J. L., Porter, W. A., Linder, J. S., and Haden, C. R., *Proc. IEEE*, 59, 323, 1971.

873. Shanefield, D., *J. Non-Crystalline Solids*, 2, 210, 1970.

874. Iizima, S., Sugi, M., Kikuchi, M., and Tanaka, K., *Solid State Commun.*, 8, 153, 1970.

875. Phillips, S. V., Booth, R. E., and McMillan, P. W., *J. Non-Crystalline Solids*, 4, 510, 1970.

876. Csillag, A., *J. Non-Crystalline Solids*, 4, 518, 1970.

877. Croitoru, N., Vescan, L., Popescu, C., and Lazarescu, M., *J. Non-Crystalline Solids*, 4, 493, 1970.

878. Feldman, D. and Moorjani, K., *J. Non-Crystalline Solids*, 2, 82, 1970.

879. Fulenwider, J. E. and Hershkowitz, G. J., *Phys. Rev. Lett.*, 25, 292, 1970.

880. Newkirk, T. F. and Hedden, G. D., *Bull. Amer. Ceram. Soc.*, 41, 281, 1962.

881. Bruyere, J. C. and Chakraverty, B. K., *Appl. Phys. Lett.*, 16, 40, 1970.

882. Hed, A. Z. and Freud, P. J., *J. Non-Crystalline Solids*, 2, 28, 1970.

883. Christopher, J. E., Coleman, R. V., Isrin, A., and Morris, R. C., *Phys. Rev.*, 172, 485, 1968.

884. Morris, R. C., Christopher, J. E., and Coleman, R. V., *Phys. Rev.*, 184, 565, 1969.

885. Lipsicas, M., Mattis, D. C., Lord, E., and Kornblitt, A., *J. Non-Crystalline Solids*, 2, 550, 1970.

886. Van Steensel, K., van de Burg, F., and Kooy, C., *Philips Res. Rep.*, 22, 170, 1967.

887. Kennedy, T. N. and Collins, F. M., Tech. Rep. No. 2, Contract No. N00014-67-A-0117-0005, 1969, unpublished.

888. Lee, P. A., Said, G., and Davis, R., *Solid State Commun*, 7, 1359, 1969.

889. Audzionis, A. I., Grigas, I. P., and Karpus, A. S., *Sov. Phys.-Solid State*, 12, 115, 1970.

890. Bullock, D. C. and Epstein, D. J., *Appl. Phys. Lett.*, 17, 199, 1970.

891. Zhdan, A. G., Sandomyrskii, V. B., Elinson, M. I., and Tchugunova, M. E., *J. Vac. Sci. Tech.*, 6, 929, 1969.

892. Szymanski, A., Larson, D. C., and Labes, M. M., *Appl. Phys. Lett.*, 14, 88, 1969.

893. Busch, G., Guntherodt, H. J., Kunzi, H. V., and Schweiger, A., to be published.

894. Hovel, H. J., *Appl. Phys. Lett.*, 17, 141, 1970.

895. Sunshine, R. A., IEEE Electron Devices Meeting, Washington, D. C., 1970.

896. Kolomiets, B. T., Lebedev, E. A., and Taksami, I. A., *Sov. Phys.-Semicond.,* 3, 621, 1969.

897. Calella, P., DeFeo, S., Hall, J., Walsh, P., Nicolaides, R., and Doremus, W., Engineering Sciences Laboratory Inform. Rep. 451, Picatinny Arsenal, 1969, unpublished.

898. Mathur, B. P. and Arntz, F. O., *J. Non-Crystalline Solids*, to be published.

899. Chopra, K. L., *Physics of Electronic Ceramics,* Hench, L. L., and Dove, D. B., Eds., Marcel Dekker, New York, 1971, in press.

900. Vogel, R. and Walsh, P. J., *Appl. Phys. Lett.*, 14, 216, 1969.

901. Shaw, R. F., *Bull. Amer. Phys. Soc.*, 15, 245, 1970.

902. Rockstad, H. K., *J. Appl. Phys.*, 42, 1159, 1971.

903. Ridley, B. K., *Proc. Phys. Soc.*, (London) 81, 996, 1963.

904. Adler, D. and Brooks, H., *Phys. Rev.*, 155, 826, 1967.

905. Hindley, N. K., *J. Non-Crystalline Solids*, 5, 31, 1970.

906. Mott, N. F., *Phil. Mag.*, in press.

907. Drake, C. F., Scanlan, I. F., and Engl, A., *Phys. Status Solidi*, 32, 193, 2963.

908. Drake, C. F. and Scanlan, I. F., *J. Non-Crystalline Solids*, 4, 469, 1970.

909. Drake, C. F., Summer School on Electrical Properties of Non-Crystalline Materials, Cambridge, 1970, unpublished.

910. Hed, A. Z. and Freud, P. J., *J. Non-Crystalline Solids*, 2, 484, 1970.

911. Bagley, B. G., *Solid State Commun.*, 8, 345, 1970.

912. Mattis, D. C., *Phys. Rev. Lett.*, 22, 936, 1969.

913. Balberg, I., *Appl. Phys. Lett.* 16, 491 1970.

914. Gunthersdorfer, M., to be published.

915. Holstrom, R., *Proc. IEEE*, 57, 1451, 1969.

916. Burton, P. and Brander, R. W., *Int. J. Electronics*, 27, 517, 1969.

917. Warren, A. C., *Electronics Lett.*, 5, 461, 1969.

918. Warren. A. C., *Electronics Lett.*, 5, 609, 1969.

919. Thomas, D. L. and Warren, A. C., *Electronics Lett.*, 6, 62, 1970.

920. Warren, A. C., *J. Non-Crystalline Solids*, 4, 613, 1970.

921. Chen, H. S., and Wang, T. T., *Phys. Status Solidi*, (a) 2, 79, 1970.

922. Cook, E. L., *J. Appl. Phys.*, 41, 551, 1970.

923. Croitoru, N. and Popescu, C., *Phys. Status Solidi*, (a) 3, 1047, 1970.

924. Sheng, W. W. and Westgate, C. R., *Solid State Commun.*, 9, 387, 1971.

925. Phillips, J. C., *Comments on Solid State Physics*, 3, 105, 1971.

926. VanMarum, M., *Ann. Phys.*, (Leipzig) 1, 68, 1799.

927. Fourier, J. B. J., *Theorie Analytique de la Chaleur*, Gauthier-Villars, Paris, 1842.

928. Joule, J. P. and Kelvin, W. T., *Free Expansion of Gases*, Ames, J. S., Ed., Scientific Memoirs No. 1, 1898, 33.

929. Von Karman, T., *Arch. Elektrotech.*, 13, 174, 1924.

930. Lueder, H. and Spenke, E., *Phys. Zeit.*, 36, 766, 1935.

931. Becker, R., *Arch. Elektrotech.*, 30, 411, 1936.

932. Fock, V., *Arch. Elektrotech.*, 19, 71, 1928.

933. Whitehead, S., *Phil. Mag.*, 37, 276, 1939.

934. Franz, W., *Encyclopedia of Physics*, Vol. 17, Springer-Verlag, Berlin, 1956, 155.

935. Dohler, G., *Phys. Status Solidi*, (a) 1, 125, 1970.

936. Sokolowski, E. J., S. B. Thesis, M. I. T., 1971, unpublished.

937. Keyes, R. W., *Comments on Solid State Physics,* in press.

938. Stocker, H. J., Barlow, C. A., and Weirauch, D. F., *J. Non-Crystalline Solids*, 4, 523, 1970.

939. Boer, K. W. and Ovshinsky, S. R., *J. Appl. Phys.*, 41, 2675, 1970.

940. Boer, K. W., Dohler, G., and Ovshinsky, S. R., *J. Non-Crystalline Solids*, 4, 573, 1970.

941. Boer, K. W., *Phys. Status Solidi*, (a) 2, 817, 1970.

942. Fritzsche, H. and Ovshinsky, S. R., *J. Non-Crystalline Solids*, 4, 464, 1970.

943. Collins, F. M., *J. Non-Crystalline Solids*, 2, 496, 1970.

944. Kaplan, T. and Daler, D., *Appl. Phys. Lett.,* in press.

945. Kolomiets, B. T., Lebedev, E. A., and Taksami, I. A., *Sov. Phys.-Semicond.*, 3, 267, 1969.

946. Weirauch, D. F., *Appl. Phys. Lett.*, 16, 72, 1970.

947. Male, J. C., *Electronics Lett.*, 6, 91, 1970.

948. Haberland, D. R., *Solid-State Electronics*, 13, 207, 1970.

949. Kanarek, K. B., S. M. Thesis, M. I. T., 1971, unpublished.

950. Henisch, H. K., Ovshinsky, S. R., and Pryor, R. W., *Proc. Congres. Int. sur les Couches Minces, Cannes, 1970*, S.F.I.T.V., Paris, 1970, 437.

951. Pryor, R. W. and Henisch, H. K., *Appl. Phys. Lett.*, 18, 324, 1971.

952. Henisch, H. K. and Pryor, R. W., *Solid-State Electronics*, in press.

953. Fritzsche, H. and Ovshinsky, S. R., *J. Non-Crystalline Solids*, 2, 393, 1970.

954. Henisch, H. K., Fagen, E. A., and Ovshinsky, S. R., *J. Non-Crystalline Solids*, 4, 538, 1970.

955. Schmidlin, F. W., *Phys. Rev.*, B1, 1583, 1970.

956. Iida, M. and Hamada, A., *Jap. J. Appl. Phys.*, 10, 224, 1971.

957. Lucas, I., to be published.

958. Lampert, M. A., *R. C. A. Rev.*, 20, 682, 1959.

959. Shanks, R. R., *J. Non-Crystalline Solids*, 2, 504, 1970.

960. Bube, R. H., *Photoconductivity of Solids*, John Wiley & Sons, New York, 1967.

961. Tyler, W. W., *Phys. Rev.*, 96, 226, 1954.

962. Lebedev, A. A., Stafeev, V. I., and Tuchkevich, V. M., *Sov. Phys.-Technical Phys.*, 1, 2071, 1957.

963. Barnett, A. M. and Jensen, H. A., *Appl. Phys. Lett.*, 12, 341, 1968.

964. Homma, K., *Electr. Commun. Jap.*, 51, 47, 1968.

965. Burnett, A. M., *Semiconductors and Semimetals*, Vol. 6, Willardson, R. K., and Beer, A. C., Eds., Academic Press, New York, 1970.

966. Chakraverty, B. K., *J. Non-Crystalline Solids*, 3, 317, 1970.

967 Homma, K., *Appl. Phys. Lett.*, 18, 198, 1971.

968. Kreynina, G. S., Selivanov, L. N., and Shomskaia, I. I., *Radio Engineering Electronics*, 5, No. 8, 219, 1960.

969. Kreynina, G. S., *Radio Engineering Electronic Phys.*, 7, 166, 1962.

970. Kreynina, G. S., *Radio Engineering Electronic Phys.*, 7, 1949, 1962.

971. Hickmott, T. W., *J. Appl. Phys.*, 33, 2669, 1962.

972. Hickmott, T. W., *J. Appl. Phys.*, 35, 2118, 1964.

973. Hickmott, T. W., *J. Appl. Phys.*, 36, 1885, 1965.

974. Simmons, J. G. and Verderber, R. R., *Proc. Roy. Soc.*, (London) A301, 77, 1967.

975. Simmons, J. G. and Verderber, R. R., *Radio Electronic Engineer*, 34, 81, 1967.

976. Verderber, R. R., Simmons, J. G., and Eales, B., *Phil. Mag.*, 16, 1049, 1967.

977. Barriac, C., Giraud-Hiraud, F., Pinard, P., and Davoine, F., *C. R. Acad. Sci.*, (Paris) 262, 423, 1966.

978. Barriac, C., Pinard, P., and Davoine, F., *C. R. Acad. Sci.*, (Paris) 266, 423, 1968.

979. Barriac, C., Pinard, P., and Davoine, F., *Phys. Status Solidi*, 34, 621, 1969.

980. Dearnaley, G., *Phys. Lett.*, 25A, 760, 1967.

981. Dearnaley, G., *Thin Solid Films*, 3, 161, 1969.

982. Dearnaley, G., Morgan, D. V., and Stoneham, A. M., *J. Non-Crystalline Solids*, 4, 593, 1970.

983. Kikuchi, M. and Iizima, S., *Appl. Phys. Lett.*, 15, 323, 1969.

984. Tanaka, K., Iizima, S., Sugi, M., and Kikuchi, M., *Solid State Commun.*, 8, 75, 1970.

985. Hiatt, W. R. and Hickmott, T. W., *Appl. Phys. Lett.*, 6, 106, 1965.

986. Argall, F., *Electronics Lett.*, 2, 282, 1966.

987. Lamb, D. R. and Rundle, P. C., *Brit. J. Appl. Phys.*, 18, 29, 1967.

988. Argall, F., *Solid-State Electronics*, 11, 535, 1968.

989. Park, K. C. and Basavaiah, S., *J. Non-Crystalline Solids*, 2, 284, 1970.

990. Sliva, P. O., Dir, G., and Griffths, C., *J. Non-Crystalline Solids*, 2, 316, 1970.

991. Chapman, R., *Electronics Lett.*, 5, 246, 1969.

992. Fritzsche, H. and Ovshinsky, S. R., *J. Non-Crystalline Solids*, 2, 148, 1970.

993. Shappirio, J. R., Eckart, D. W., and Cook, C. F., *J. Non-Crystalline Solids*, 2, 217, 1970.

994. Bagley, B. G. and Northover, W. R., *J. Non-Crystalline Solids*, 2, 161, 1970.

995. Bagley, B. G. and Bair, H. E., *J. Non-Crystalline Solids*, 2, 155, 1970.

996. Phillips, S. V., Booth, R. E., and MacMillan, P. W., *Proc. Brit. Ceram. Soc.*, 18, 1, 1970.

997. Roy, R. and Caslavska, V., *Solid State Commun.*, 7, 1467, 1969.

998. Tanaka, K., Iizima, S., Sugi, M., Okada, Y., and Kikuchi, M., *Solid State Commun.*, 8, 1333, 1970.

999. Uttecht, R., Stevenson, H., Sie, C. H., Griener, J. D., and Raghavan, K. S., *J. Non-Crystalline Solids*, 2, 358, 1970.

1000. Bunton, G. V., *J. Non-Crystalline Solids*, 6, 72, 1971.

1001. Sample, H. H., Gerber, J. A., Neuringer, L. J., and Kaufman, L. A., *Phys. Lett.*, 33A, 119, 1970.

1002. Vermilyea, D. A., *J. Electrochem. Soc.*, 102, 207, 1955.

1003. Feinleib, J. and Ovshinsky, S. R., *J. Non-Crystalline Solids*, 4, 564, 1970.

1004. Evans, E. J., Helbers, J. H., and Ovshinsky, S. R., *J. Non-Crystalline Solids*, 2, 334, 1970.

1005. Feinleib, J., *Int. J. Magnetism*, in press.

1006. Feinleib, J., de Neufville, J., Moss, S. C., and Ovshinsky, S. R., *Appl. Phys. Lett.*, 18, 254, 1971.

1007. Adler, D. and Feinleib, J., *Physics of Opto-Electronic Materials*, Albers, W. A., Ed., Plenum Press, New York, 1971, 233.